PORCHES, DECKS & OUTBUILDINGS

THE BEST OF
Fine Homebuilding

PORCHES, DECKS & OUTBUILDINGS

THE BEST OF
Fine Homebuilding

The Taunton Press

Cover photo: Charles Miller

Back-cover photos: Lynn Karlin (top), Charles Miller (center), Jeff Kolle (bottom)

BOOKS & VIDEOS

for fellow enthusiasts

First printing: 1997
Printed in the United States of America

A Fine Homebuilding Book

Fine Homebuilding® is a trademark of The Taunton Press, Inc.,
registered in the U.S. Patent and Trademark Office.

The Taunton Press, Inc.
63 South Main Street
P.O. Box 5506
Newtown, Connecticut 06470-5506

Library of Congress Cataloging-in-Publication Data

Porches, decks & outbuildings : the best of Fine homebuilding.
    p. cm.
    Includes index.
    ISBN 1-56158-181-X
    1. Porches Design and construction. 2. Decks (Architecture,
Domestic)—Design and construction. 3. Garden structures—Design and
construction. I. Taunton Press. II. Fine homebuilding.
    TH4970.P675   1997
    690'.184—dc20
                                          96-43023
                                            CIP

CONTENTS

(continued)

INTRODUCTION

A WHILE BACK I VISITED a friend's new house out in Bellingham, Washington. It's a Victorian farmhouse with a covered porch about 8 ft. deep running all the way around it. The porch is only a couple of steps up from grade, so no handrails are necessary, and there are no railings to funnel you in one direction or another.

You can step up to that porch from anywhere you want. That porch is like a big pair of arms welcoming you into the home's embrace. Despite the lack of railings, there are plenty of broad steps to sit on and sturdy posts to lean against.

I keep thinking about that porch and the way it made me feel. I'm pretty sure that at the end of a long, hard day it would be a good place to sit and enjoy a cold beer. And I think I've finally figured out a way to incorporate a porch like that into the new addition I'm building on my own house.

Whether it's a front porch, a screen porch, or a backyard deck, every home ought to have an outdoor space where you can relax at the end of the day or sit and watch the rain.

But exposed to weather like they are, porches and decks can be tricky to build. The right materials are critical, as are construction details that will shed water. In this book, which is a collection of articles from *Fine Homebuilding* magazine, you'll find advice on materials and details, along with design ideas for all sorts of porches, decks, and outbuildings. Written by builders and architects, who are discussing their own projects, these articles are the voice of experience.

–Kevin Ireton, editor

The American Porch

A look at porches, and a glossary of types from past to present

by Davida Rochlin

When I was a child my family would take yearly vacations to the border town of Nogales, Ariz., to visit my grandmother. She lived in a brick house that sat on top of a hill. There were no other houses in sight for at least 20 miles. The Arizona-Mexico crossing station was a vague outline in the distance.

Her home had a screened porch, L-shaped in plan, that bent around the southeast corner. During the hot desert summers, it was the porch that provided protection from the afternoon thunderstorms, and refuge at night from the sweltering heat. Being a large group of grandchildren, sisters, brothers and cousins, we were regularly shuttled off to the porch to play, still within adult earshot, but out of the way.

I was raised in Southern California, and our two-story 1920s bungalow, formerly occupied by singing cowboy Hoot Gibson, sported a wealth of porch space. The veranda, side porch, sunroom and sleeping porch were all viewed as hideaways in a household that rocked with constant noise and activity.

The porches were just there, sociable and open—unassigned parts of the house that belonged to everyone and no one. They were shady transitions between indoors and outdoors

***Victorian extravagance.* Designers and builders of Victorian homes had more fun with porches than anybody before or since. The time was right, since materials and tools were available to do intricate work, and the homes' inhabitants were still interested in spending time between the sidewalk and the house, engaging the members of the neighborhood. This house is in Cape May, N. J.**

where one could sip iced tea with a neighbor, share a summer supper or sneak away to read a book. The informal atmosphere encouraged conversations and the easy relationships that were such a part of family life.

In America, the front porch has played a significant social and cultural role. In its most fundamental role, it served as shelter from the elements. But from this the porch evolved into an outdoor living room, an adjunct that offered the personal security of the home with all of nature's cooling advantages.

Porches from the past—The word porch comes from the Greek *portico*, a roof with classic columns. It is now defined as a covered entrance to a building. As settlers came to the United States, they brought building styles from

The evolution of the porch in America

The porch is a building form that has worked its way backwards into our building vocabulary, from the mighty to the more humble and accessible. It began in classical Greece as a portico—a row of columns supporting a roof adjacent to a temple or other public building. In time, the word came to mean any quasi indoor/outdoor space that was sheltered by a roof, but without walls. Today's distant kinfolk of the portico include solar greenhouses, trellises and the ubiquitous barbecue deck. The drawing below shows the plans of the most common porch styles.

Front porches

Portico: a roof with classic columns.
Stoop: of Dutch descent, a covered porch with seats at each side of the house door.
Vestibule: of English descent, a small entrance hall or room either to a building or to a room within a building.
Veranda: the place of sociability. Differs from the porch or gallery in that the depth of the veranda floor is wider than that of a porch. It extends the length of two sides of the house or encircles the entire building.
Colossal portico: a colonnade rising the full height of the building.
Colossal portico with insert balcony: a colonnade rising the full height of the building with a balcony over the entry.

Living porch: an outdoor living room.
Double deck: a two-story porch.
Gallery: a covered porch open at one side.
Two-story gallery: a two-story porch open at one side.
Open porch: a small entrance porch.
Arbor porch: a trellis with planted vine.
Arbor veranda: a trellis with planted vines encircling the entire house.
Screened porch: an indoor-outdoor room that is screened to protect against mosquitos and other insects.
Piazza: of Italian descent. A porch often associated with the Victorian era.

Appendages

Deck: an outdoor platform, usually without a roof. The modern alternative to the porch.
Balcony: an economical, contemporary version of the porch. Commonly used in apartment complexes to give dwellers an outdoor living space.
Pavilion: a partially enclosed platform, usually with a roof.
Terrace: similar to a patio, a place of privacy connected to the outdoors.

Detached semi-structures

Colonnade: a series of columns set at regular intervals.
Trellis: a structure of wooden or metal strips on which vines are trained.

Gazebo-pergola: a freestanding structure usually designed for a garden.

Special-use porches

Back porch: implied use is for service facilities. The back porch has been used for washing up, morning shaves, and a place to prepare meals.
Side porch: popularized by 1850s pattern books as an adjunct to other porches.
Sunroom: acts as a greenhouse or place to keep plants.
Breakfast nook: usually adjacent to the kitchen. A morning room; eastern exposure is suggested.
Sleeping porch: descendant of the loggia. Usually on the second floor in the back of the house. Two or three sides of the sleeping room are open air.
Roof deck: an outdoor living space on the roof of a house or apartment. A convenient place to sunbathe.
Kitchen porch: a utility area adjacent to the kitchen. Used as a storage place for mops, brooms etc.
Awning: a cloth shade used as protection from the rain or sun.
Dog trot: indigenous to the United States, a hallway or central porch that runs from one side of the house to the other.
Loggia: a roofed, open gallery.
—D. R.

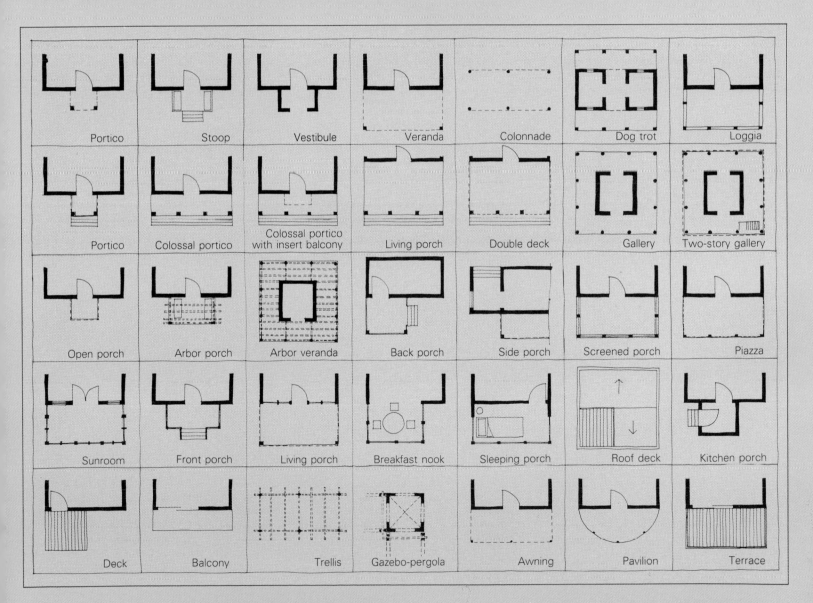

Drawing: Davida Rochlin

their homelands. Melting-pot antecedents of the porch include the Dutch stoop, English vestibule, Roman colonnade, Greek loggia, and Indian veranda. Today, there are over 30 readily recognizable offshoots of the original form (see the facing page).

The dog-trot is one of the few indigenous porches in the United States. Named after the dog that trotted back and forth, it was a breezeway that separated two portions of the house. People left the doors open on either side of the dog-trot house to promote ventilation. Summer breezes made the dog-trot the coolest part of the house, and it became the principal area for cooking, child care and homework. Dog-trot porches were first seen in log-cabin structures in the Appalachian mountain regions. This simple breezeway has been integrated by contemporary architects into modern residential design (see *FHB* #28, pp. 73-77). Many homes retain the concept of the dog-trot breezeway, where the dog-trot marks the entry leading to a courtyard patio.

In the South, the porch first served as protection from the warm weather. But eventually, southern porches evolved into settings for family life that integrated indoor and outdoor spaces into stylish new structures. The largest, the colossal portico, is a porch that runs the full height and length of the house. The colossal portico on the front of George Washington's Mount Vernon exemplifies this style, which was influenced by the 16th-century Italian architect, Andrea Palladio. Others expanded on the idea by adding a second-story balcony, creating the colossal portico with insert balcony (photo top right). Second-story insert and balcony porches were used to define entryways.

The gallery, a porch around all four sides of a house, was used extensively in New Orleans to screen out the oppressive sunshine. The house and its gallery were placed atop a raised basement for protection from flooding. The two-story porch with Greek columns became the hallmark of antebellum plantation houses. Larger plantation houses had two-story porches across the front and rear facades. Unlike the front porch, the back porch did not extend the full width of the house, and it was flanked by side rooms. The back porch was used as a service and storage area, while the grand front porch was used as a place to greet guests or sit before mealtime. Often, two-story galleries encircled the entire house. In Charleston, S. C., such houses were built with narrow ends facing the street. The south and west sides opened onto gardens. The galleries with garden views provided shade during the summer and allowed southern light to enter during the winter.

In the latter half of the 19th century, ornamental cast-iron railings became available. Because these railings were mass produced, ready to assemble and inexpensive, balconies and porches in coastal southern cities like Savannah and New Orleans began to assume the ornate, lacy quality that has long since been associated with the deep South (photo middle right).

In the same period, architectural pattern books encouraged Americans living in the countryside to build porches onto their homes. Porch types explained in these books included the open porch, a trellis-topped entry that acted as a partial shelter, an arbor veranda and a trellis covered with grapevines that wrapped around the entire house. For inside the house, the pattern books suggested a sunroom, a glassed-in garden room, a breakfast nook or a porch adjacent to the kitchen (often facing east for morning sun), and the back porch, used exclusively to hang laundry or dry a mop.

The 19th century saw the flowering of Victorian extravagance as Queen Anne, Eastlake and Italianate styles bloomed. Porches became crowded with rocking chairs, wicker settees, hammocks and gliders. People began to pay more attention to architectural detailing, and the use of latticework, bric-a-brac and railing on verandas became widespread (photo p. 8). This resulted in contrasts of solid and void, light and dark, open and closed.

Porches at the turn of the century—First appearing in California in the early 1900s, the bungalow-style house was popularized in the

Two Southern porches. **A combination veranda with colossal portico, top right, encircles this house in Georgia. An insert balcony above the front door provides another place to sit outdoors, and emphasizes the placement of the entryway. In the late 19th century, cast-iron components became available to builders, and lacy railings and balustrades became a Southern signature. Some multi-storied apartments include a balcony as a place to get out of the sun and enjoy the evening breezes, such as the building at right, in New Orleans, La.**

Inside and out. **A porch is a transition between the street and the house. Inside the home, tables and chairs occupy formal positions, and the temperature can be controlled. On the porch, things are more flexible. Rocking chairs and gliders are the typical furnishings, and the temperature is subject to the breeze off the lawn. The porch below is in Oak Park, Ill.**

Top photo: Georgia Bureau of Tourism; Other photos: Davida Rochlin

pattern books of the time. It was the last type of vernacular housing to feature a full front porch. Bungalow porches (photo top left) were usually topped with a gable roof that was supported by brick or fieldstone columns. These generous front porches provided a link between the community and the house.

Along with the new century came an emphasis on health, fresh air and the vigorous life. To provide the fresh air, many a bungalow included a sleeping porch placed at the back of the house for quiet and privacy. This amenity reached its zenith in the Greene brothers' Gamble house, in Pasadena, Calif. (middle photo, far left). The Gamble house sleeping porches were placed on the second floor, where their occupants enjoyed the serene vistas of the nearby San Gabriel mountains.

The decline of the porch—In the mid-20th century, porch life went into decline. Automobile pollution and noise made facing the street unpleasant, and television replaced porch sitting as the dominant form of passive entertainment. By the 1960s, air conditioning made it unnecessary to cool off outdoors. With more and more distractions competing for attention, city dwellers began to value their privacy, and the focus of the suburban house shifted to the backyard, with its pool and patio.

Modern houses have porch-like appendages that are reminiscent of the porches of the past. Probably the most common one is the deck, followed by terraces and balconies (middle photo, near left). Some passive-solar homes have a glassed-in greenhouse designed to catch the sun, which also provides a space to grow flowers and vegetables all year long. On a smaller scale, many front porches have now become glassed-in sunrooms to take advantage of the additional square footage. These forms are still linked to nature, but the community tie is lost.

Twentieth-century porches. The last housing type to be built on a mass scale that included a front porch was the bungalow, top. The porches usually ran the length of the house. Rustic materials were often used, such as these river-rock columns. Bungalows often had sleeping porches on the second floor, facing the backyard. The sleeping porch on the Greene brothers' Gamble house, above left, represents the pinnacle of the style.

Porches in decline. With the coming of the car, porches began to disappear, re-emerging as backyard decks and upstairs balconies. This balcony in Santa Monica, Calif., above right, has done both.

Revival. Once again the porch is making its presence felt in American residential design. This house by San Francisco architect William Turnbull is encircled by a gallery that shelters the home from the summer sun. A trellised roof covers a courtyard in the center of the house.

New signs of porch life—Although it may be down, the American porch is not out. In fact, it's showing signs of making a comeback. Throughout the United States, housing developments are beginning to tout the front porch as a selling point. In Valencia, Calif., one developer is including a veranda as a major design feature of his houses. The development offers a limited backyard and reorients outdoor activity to the front porch. At Seaside, Fla., an 80-acre planned community of individual homes features screened front porches.

Contemporary custom designs are once again showcasing the porch. San Francisco architect William Turnbull frequently integrates one or more porches into his homes. One such residence is in Modesto, Calif. (photo bottom left). The weather in Modesto is hot, and this house is encircled by a covered porch that acts to cool the home. The center of the house is an open courtyard, festooned with elaborate trellises that turn it into an interior gazebo, reminding us that the style and grace of another time can still be appreciated today. □

Davida Rochlin, AIA, is an architect and porch lover living in Los Angeles.

Classical Style in a Porch Addition

Tips from a restoration expert on deck construction, column building and weatherproof design

by Ted Ewen

Those who undertake exterior renovations on a Greek Revival house are faced with a touchy and challenging venture. Changes or additions must be compatible with the original design; they must also fit in with modern day-to-day activities. Even a small porch addition like the one I built recently must fulfill these two requirements. Of course another problem with porch construction is the weather: Porches won't last long if they are built without regard for weatherproofing. Fortunately I've been able to draw on quite a bit of boatbuilding experience, using materials and design details that have stood up to marine conditions.

The new porch was to replace a large laundry room that had been built onto the south side of a Greek Revival mansion in 1928. The main house, built between 1830 and 1840, overlooks the Hudson River from a high vantage point in Scarborough, New York. The old laundry room, like the kitchen it faced, was built at a time when cooking and cleaning were done by servants on sizable estates like this one. When the new owners of the house contacted me early in 1980, we decided to tear down the old laundry room. The new porch would expand the formerly viewless kitchen with its open deck and pleasant view. It would also give the kitchen a bit more formality as a main entry into the house. My job was to build something compatible with the main house. A repeated column and arch relief on the laundry room provided some important size and scale details. Having agreed with the owners on the design and drawn up plans, I set to work undoing the work of 50 years before.

A new deck—Tearing down the laundry room was a time-consuming job. It had been extreme-ly well built—knit together by carpenters who loved nails and knew how and where to drive them. The foundation was 12-in. thick reinforced concrete; the interior walls and ceiling were 1¼-in. concrete and plaster over galvanized wire lath. Over the years I've dismantled quite a few buildings, but this one really took the cake, especially considering its diminutive size (about 12 ft. by 16 ft.).

Before taking the roof off the laundry room, I removed the wood floor and the old floor joists to expose the crawl space underneath. Enlisting the help of the owner and a friend, I regraded the crawl space, installed 6-in. diameter PVC pipe to drain it outside the walls, and then floated 2 in. of concrete over galvanized lath laid on the crawl-space floor. We also poured two new concrete piers at the center of the foundation to support the new floor joists. The old roof

Having torn down an old laundry room and built a new deck on the south side of a Greek Revival mansion, the author prepares to extend the existing girder. Once in place, this girder will carry the ceiling joists for the new porch.

was left in place to protect the foundation from rain until the concrete cured.

With its beautiful view of Haverstraw Bay and the Hudson River Valley, the new porch and its columns would be exposed to heavy weather. Since the exposed section of the old deck had suffered considerably from water damage, I knew I would have to make the new construction and surface finish as durable as possible. I used pressure-treated 2x6s for all floor joists and spaced them 12 in. on center. Hot-dipped galvanized nails and screws, liberal applications of Woodlife preservative to all lumber (especially to the end grain) and a 2½-in. pitch in the deck's 24-ft. length were some of the measures I took to improve the weather resistance of the new construction. I also selected my lumber board by board whenever I could. Not all lumberyards will allow you to pick and choose, but using only straight, clear-grained stock was an extra assurance that the porch columns and their framework would last a long time.

For the deck I used 4-in. wide, 5/4 tongue-and-groove fir, toenailing directly into the 2x6 joists with 10d common nails. I took the time to drill pilot holes in both planks and joists to avoid splitting the wood. After nailing down every three courses of planks, I allowed a space of nearly 1/16 in. between the third and fourth courses. This gives the deck room to expand when moisture swells the boards. If you've ever seen a beautifully laid plank floor buckle in humid weather, you can appreciate the need for expansion joints. The total expansion space for the width of this deck is 1½ in., and I ran a seam of butyl rubber in each joint, to keep dirt out.

Making the columns—To duplicate the relief on the square columns of the main house, I measured them and then scaled the new columns accordingly. Proportions and detailing had to be kept the same to ensure compatibility with the rest of the house. The columns on the west side of the deck would be built in three parts: a large square main column flanked by the two smaller columns on which the arches would rest (see the drawing, below right). As an aid in laying out the columns, I drew full-scale cross sections on a piece of plywood and protected these working drawings with a wash coat of var-

nish. This way, they can be used on another job or as a guide for replacing a damaged column at some future date.

To build the columns, I used clear, No. 2 white pine boards, ¾ in. thick. As shown in the drawing at right, I built the main columns with an inner and outer shell for load-bearing rigidity and general durability. The edges of the inner planks clear each other and the outer shell by about ⅛ in. The gap will permit the outer shell to shrink—as it will over the years—without having its joints forced apart by bearing on the inner planks, which will shrink much less. I fastened the inner pieces to the outer shell before putting the outer shell together, using 1¼-in. and 1½-in. #10 flathead screws.

To build up the corners and ends of each column to create a relieved panel in the center, I used a combination of ¼-in. lattice (strips of pine in various widths), marine-grade mahogany plywood (¼ in. thick) and stock moldings. The lines created by these raised strips and panels give the otherwise flat columns an appearance of lightness and fine detail. It's easy to understand why they were incorporated into the column motif on the original house.

The trim was fastened to the columns with aluminum nails and Phenoseal adhesive caulking (made by Gloucester Co., Box 428, Franklin, Mass. 02038). The adhesive caulking is a new material I discovered through my boatbuilding contacts. The waterproof bond is flexible and mildew resistant, two useful qualities as the wood ages and is exposed to moisture.

After planing and sanding the edges of the trim, I treated each column with Woodlife, waited several days for the preservative to penetrate and dry, and then brushed on a coat of Benjamin Moore white exterior alkyd primer. Then the columns were ready to go up, and construction could begin in earnest.

Column raising—Rather than rest the columns directly on the deck planks, I cut plywood pads conforming to the plan section of each column and screwed them to the deck where the columns would stand. The pads even out the downward pressure of the columns. More important, they elevate the vulnerable end grain of the column planks above the surface of the deck. Each

A full-scale template, drawn on ¼-in. plywood, serves as a guide during column construction. Here the arch post is test-fit on the half-column that will go against the side of the house.

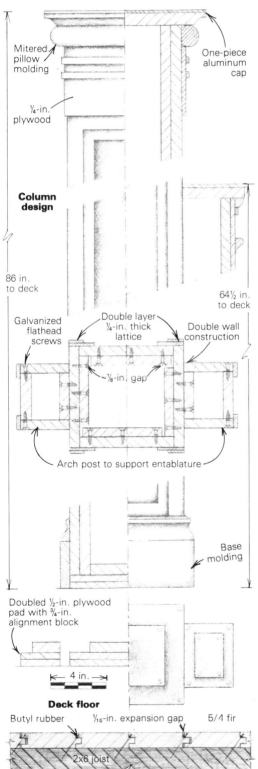

Column design

Mitered pillow molding
¼-in. plywood
One-piece aluminum cap
86 in. to deck
64½ in. to deck
Galvanized flathead screws
Double layer ¼-in. thick lattice
Double wall construction
⅛-in. gap
Arch post to support entablature
Base molding
Doubled ½-in. plywood pad with ¾-in. alignment block
4 in.

Deck floor
Butyl rubber
1/16-in. expansion gap
5/4 fir
2x6 joist

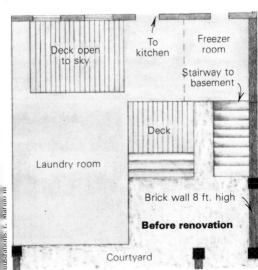

Deck open to sky
To kitchen
Freezer room
Stairway to basement
Deck
Laundry room
Brick wall 8 ft. high

Before renovation

Courtyard

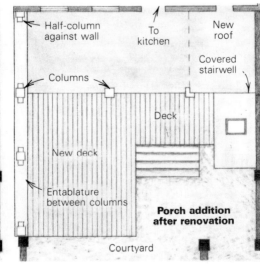

Half-column against wall
To kitchen
New roof
Columns
Covered stairwell
Deck
New deck
Entablature between columns

Porch addition after renovation

Courtyard

Test-fitting the soffit in the new roof. Joints between old and new soffit, fascia and frieze boards are staggered for lateral strength and to make the addition as inconspicuous as possible. One-by-three spacers build out the fascia to the full width of the column.

pad consists of two stacked pieces of ½-in. A/C plywood. Atop each pad I screwed a piece of ¾-in. plywood cut to match the interior dimensions of the column. This aligned the columns and also gave them extra resistance to horizontal pressures. Because of the pitch built into the deck, my square-bottomed columns would not rest plumb when raised into position. I had planned for this and constructed each column a bit longer than necessary. Using a large adjustable bevel, I transferred the pitch to the bottom of the column and then cut accordingly. The pad would be concealed by a base molding.

The procedure for column raising was as follows: I lifted the column into position on its pad, braced it temporarily with diagonal struts, installed an aluminum cap on the top of the column to prevent water infiltration, and then connected the top of the column to its neighbor with a new girder, soffit and fascia board. I had to run the girder (a 2x10 and 2x8 spiked together) all the way back to the corner of the freezer room because the existing beam had decayed badly. Then I nailed short lengths of 2x3s vertically

along the outside face of the girder to extend the width of the soffit that would cover these spacers. As for the soffit, fascia and frieze board, the joints where these new boards met their old counterparts were staggered to increase the lateral strength of the new roof section and to make the meeting of new and old less conspicuous (photo above). I vented the soffit by drilling ¾-in. dia. holes and covering them with small squares of galvanized hardware cloth.

One detail of the frieze bears mentioning. At 13½ in. from top to bottom, the frieze board had to be made from two planks, joined along one edge. Rather than use a 90° edge joint, which would allow moisture to make its way in behind the board, I cut a beveled edge joint (the detail can be seen in the drawing on the facing page). The top edge of the bottom board slants upwards, discouraging water penetration. I covered the joint with a taenia molding, completing the frieze and further increasing its resistance to the weather.

The roof itself posed no problems—I simply had to continue the line and pitch of the existing

roof that covered the freezer room. After lag-bolting a 2x10 header to the south wall, I anchored 2x6 rafters to the header with joist hangers. Each rafter had to be notched to fit the girder connecting the two columns (photo facing page, top right). Blocking (2x4s) followed between rafters for extra stability. Since the owners wanted a flat ceiling, the slanted rafters couldn't serve doubly as ceiling joists. I fastened 2x4 joists to the header and ran them level all the way across to the girder. Tapering the ends of the 2x4s allowed me to spike them directly to the rafters. I reinforced each joist-to-rafter connection by nailing plywood plates to both members. Then I nailed ⅝-in. plywood to the rafters and installed a triple-ply felt roof, with emulsion between layers and an aluminum/asphalt topcoat. The aluminum finish will reflect sunlight, thus preventing the heat build-up which shortens the life of dark-colored felt roofs.

The entablature—An arch taken from the old laundry room served as the model for the new arched entablature that would complete the classical detailing on the porch. I traced the arch on a 4x8 sheet of ¼-in. exterior-grade mahogany plywood, cut out the curve with a saber saw (photo below, center) and then cut the plywood to fit the space between columns. As shown in the drawing below, the entablature consists of two sheets of plywood held together by interior braces and a curved soffit (¼-in. plywood bent to the radius of the arch and glued to the interior braces). Once I had cut and trimmed the first piece to fit between the two columns, I used it as a template for cutting the next piece on the west side of the porch. Rather than assemble the entablature completely before installing it between the columns, I reinforced each sheet and then lifted it into position against nailing strips fastened to the large columns. With this arch glued and nailed in place, I installed its mate in the same manner (photo bottom left). Gluing and tacking the curved soffit between the twin arches was the next step. With the entablature in place, the porch was structurally complete. The arch and column motif will be repeated twice along the edge of the deck (photo, below right), creating a colonnade. The new addition is open and informal, yet very much part of the Greek Revival tradition of the rest of the house. □

Ted Ewen, of Scarsdale, N.Y., specializes in restoring and renovating historic buildings.

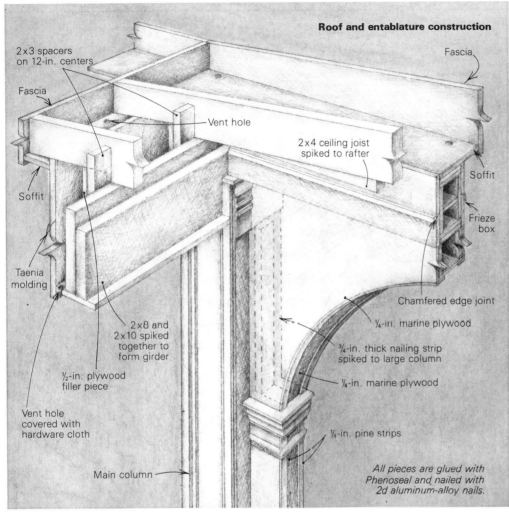

Roof and entablature construction

2x3 spacers on 12-in. centers

Fascia

Fascia

Vent hole

2x4 ceiling joist spiked to rafter

Soffit

Soffit

Frieze box

Taenia molding

2x8 and 2x10 spiked together to form girder

Chamfered edge joint

¼-in. marine plywood

¾-in. thick nailing strip spiked to large column

½-in. plywood filler piece

¼-in. marine plywood

Vent hole covered with hardware cloth

¼-in. pine strips

Main column

All pieces are glued with Phenoseal and nailed with 2d aluminum-alloy nails.

Rafters for the new roof are supported on one side by a 2x10 header bolted to the house, on the other by a girder between columns. Blocking between rafters adds rigidity.

Using an arch detail from the old laundry room as a model, the author cuts out the arch that will be fit between columns.

Nailing strips are fastened to the ceiling and new column sides (left) in preparation for attaching one plywood face of the entablature. Right, two identical plywood arches, fastened to nailing strips and joined with a curved soffit, complete the entablature between columns. Repeating this motif will connect the two additional columns on the finished porch, making a colonnade.

Porch Ornamentation

A sampling from the well-preserved
19th-century seaside resort town of Cape May

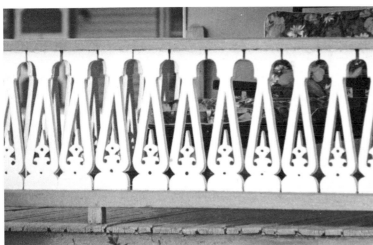

Although Cape May, N.J., had been a busy resort since the advent of the steamboat, it really made a bid for popularity in 1852, when construction began on the 3,000-room Mount Vernon Hotel, billed as "the largest in the world." Two years in the building, the grand hotel was destroyed in 1856, in one of the many fires in the town. But this imposing symmetrical structure, with its delicate porches and balconies, set a standard for the residential expansion that followed the opening of the Cape May Railroad in 1873.

The simple, lightly ornamented rectangular buildings of 1850 to 1860 were elaborated on, notably by Stephen D. Button, a Philadelphia architect who designed more than 40 houses in Cape May from 1863 to 1893. Details from three of these are shown at far left. At top is a detail from a cottage on Stockton Place, a row developed from 1871 to 1872. The scrollsawn spandrels (embellishments between columns) are typical of this period. Carved spandrels were used on the John B. McCreary residence (1869), below right, a Victorian Gothic atypical of Button's work. The acroterion (peak ornament), below left, was also in favor. Most often these were of wood, elaborately scrollsawn, but cast iron sometimes served.

The scrollsaw allowed for an endless variety of patterns of dark and light on balustrades, in brackets and pierced panels; flat lattice strips were also used to good effect by varying their orientation. The two double-tiered buildings at top (left and right) are typical of the rebuilding that was done after the 1878 fire, in the style of the 60s and 70s but on a more intimate scale. The cottage at center, by an unknown architect, was built in 1881, based perhaps on Sloan's *Model Architect*, a popular pattern book of this time. The delicate vergeboard gives it a modest distinction. For more on Cape May, see *Cape May, Queen of the Seaside Resorts,* by G. Thomas and C. Doebley (Associated University Presses, 1976). —Betsy Levine

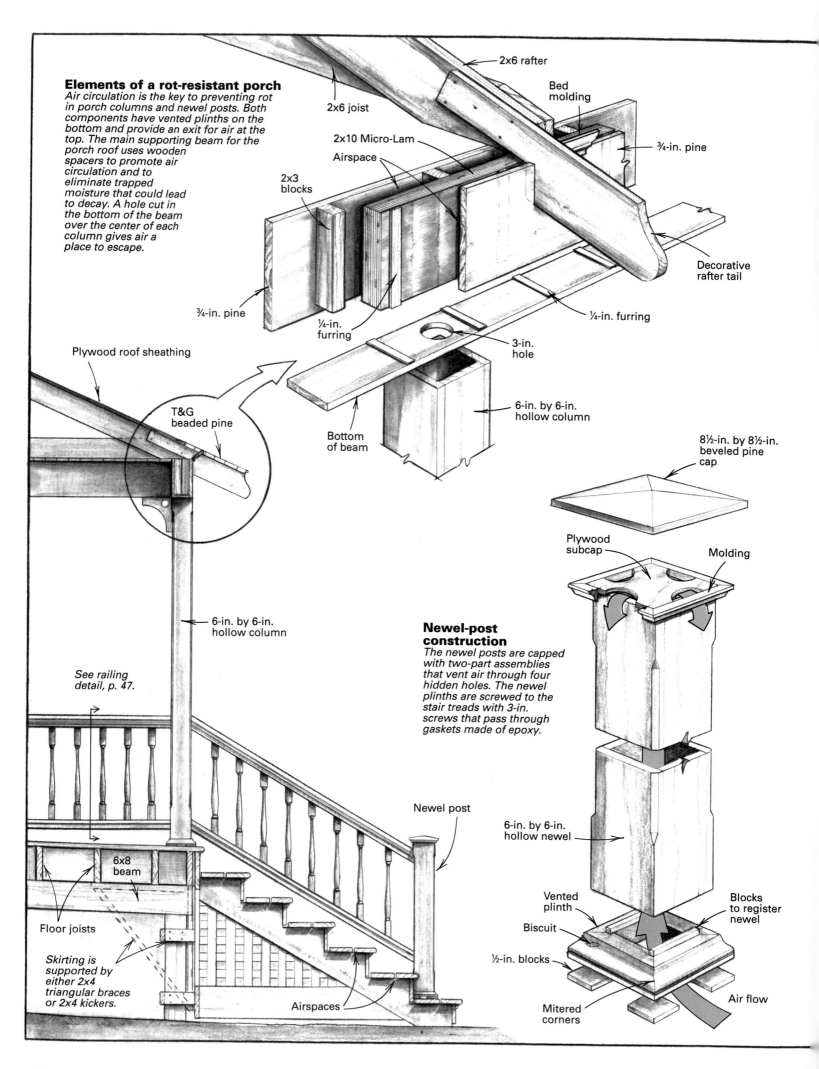

Elements of a rot-resistant porch

Air circulation is the key to preventing rot in porch columns and newel posts. Both components have vented plinths on the bottom and provide an exit for air at the top. The main supporting beam for the porch roof uses wooden spacers to promote air circulation and to eliminate trapped moisture that could lead to decay. A hole cut in the bottom of the beam over the center of each column gives air a place to escape.

2x6 rafter

2x6 joist

Bed molding

¾-in. pine

2x10 Micro-Lam

Airspace

2x3 blocks

Decorative rafter tail

¾-in. pine

¼-in. furring

¼-in. furring

3-in. hole

Plywood roof sheathing

T&G beaded pine

Bottom of beam

6-in. by 6-in. hollow column

8½-in. by 8½-in. beveled pine cap

Plywood subcap

Molding

6-in. by 6-in. hollow column

Newel-post construction

The newel posts are capped with two-part assemblies that vent air through four hidden holes. The newel plinths are screwed to the stair treads with 3-in. screws that pass through gaskets made of epoxy.

See railing detail, p. 47.

Newel post

6-in. by 6-in. hollow newel

6x8 beam

Floor joists

Vented plinth

Biscuit

Blocks to register newel

Skirting is supported by either 2x4 triangular braces or 2x4 kickers.

Airspaces

½-in. blocks

Mitered corners

Air flow

Drawings: Bob Goodfellow

Porches That Won't Rot

For a long, healthy life, these structures need a breath of fresh air

by Kevin M. Mahoney

Even as a boy, I'd happily go out of my way to look at a good porch. I'd ride my bike down the streets of old neighborhoods in Buffalo, New York, and imagine myself living in houses that I passed. My favorites were the houses with distinctive front porches. Massive and ornate, or simple and elegant, the porch made the house.

I'm not surprised that I now work as a carpenter for a company specializing in building and restoring Victorian front porches. During the summer we spend more than half our time either building new porches or fixing old ones. I think Garrison Keillor got it right when he described a good porch as a place that "lets you smoke, talk loud, eat with your fingers... without running away from home."

Because these open-air structures add so much to a house, I'd like the ones I build to last forever. That isn't literally possible. But we have developed a system for wood-porch construction that makes great strides toward that end. As we worked on older porches, it became obvious that what hurt them most was trapped moisture and lack of air circulation. That's what destroys columns, floors and framing. Some builders of an earlier era avoided these problems by using hollow posts and beams and by finding other ways to circulate air through the structure. We combine some of these time-honored techniques with a few of our own, and we add the

RULES OF THUMB

Our approach to porch construction can be boiled down to a handful of general rules that will help any wooden structure survive outdoors.

- **Use pressure-treated lumber for deck and stair framing and for all components that come into contact with the ground.**

- **Encourage air flow beneath the porch and the stairs.**

- **Pitch the floor away from the house to allow water runoff.**

- **Seal all end grain.**

- **Vent columns and newel posts at both top and bottom.**

- **Avoid unnecessary wood-to-wood contact that can trap moisture.**

Reconstruction nearly complete. At this Buffalo, N. Y., Victorian house, the front porch was rebuilt using techniques designed to eliminate trapped moisture that leads to rot.

Nonskid stair treads. **A strip along the front of each stair tread gets a coating of epoxy and crushed walnut shells for good footing in poor weather. The furring at the top of the stairs will separate the top riser from the board beneath it and prevent trapped water that could rot the structure.**

protection that modern paints and sealants can offer. Our oldest projects, going back 11 years, still show no sign of decay, so we think we're on the right track.

Failure in the usual places—We recently rebuilt a porch for Robert and Denise Sheig in Buffalo (photo p. 19). Their house is a beautiful turn-of-the-century Victorian that sits on a tree-lined street in the city's Delaware Park district. Parts of the original porch have held up for nearly 100 years; parts that failed were typical trouble spots. The wooden stairs and railings had long since deteriorated and had been replaced in the 1940s by concrete stairs and iron railings. The flooring below the columns, and the bottoms of the columns themselves, had rotted, as well as the spindle ends in the porch railing. The porch skirting had decayed where it pressed against the ground. And the porch beam and the exposed rafter tails had decayed where either the roof or the gutter built into the roof (called a Yankee gutter) had leaked.

Our first task was to rebuild the supporting posts and beams and the deck framing. After bracing the main roof beam from the ground, we removed old columns, railings, concrete stairs and decking. About 50% of the existing floor framing could be salvaged; the rest had rotted beyond repair.

The floor joists rested on three 6x10 beams, 8 ft. o. c., running perpendicular to the house. The 6x10 beams, which slope ¼ in. per ft. so that water drains away from the house, were supported by rusting metal posts. After bracing the beams we removed the metal posts; then we dug three 48-in. deep holes and poured in about 8 in. of concrete for footings. After the concrete had set, we added 3 in. of gravel for drainage, coated the new 6x6 pressure-treated posts with Benjamin

Moore Moorwood Penetrating Clearwood Sealer and set the posts on top of the gravel in the holes. We topped off each hole with concrete and brought in some fill to pitch the grade under the porch away from the house.

The 2x10 floor joists, set 16 in. o. c., sit on top of the beams and are parallel to the house. We replaced the rotted joists with pressure-treated lumber. Normally, a 2x rim joist would be nailed to the ends of the joists to strengthen and stabilize the frame. But eliminating extra layers of wood reduces the chance of decay, so instead we used a 1x12 pine apron to cap the joist ends. The 1x12 serves the same purpose as a rim joist and also becomes the top rail of the skirting that we built later. This eliminated a layer of wood that might later hold moisture and lead to rot.

Along the front of the porch, the 1x10 pine apron had to be nailed to the face of the last joist. That couldn't be avoided, but we used ¼-in. thick furring run vertically 12 in. o. c. to separate the apron from its neighboring joist. We use the same technique wherever possible: When two boards must be face nailed, we separate the two with blocks or furring to encourage air circulation.

Deck must shed water—Once the deck framing was complete, we allowed it to set for about three weeks before installing the deck boards. Pressure-treated lumber shrinks during this period, and decks can buckle if they are installed on new framing too soon. The 5/4 T&G flooring, 3 in. wide, runs perpendicular to the house so that water won't become trapped between boards as it runs off the deck. We painted the tongues and grooves with Benjamin Moore Alkyd Urethane Reinforced Porch and Floor Enamel and installed the decking while the paint was still wet. Painting the tongues and grooves is essential, but there is no reason to lose time waiting for those

parts to dry. The paint provides a good bond between boards. We also coat the tops of all floor joists with a penetrating sealer before the deck boards are installed.

Each deck board was blind nailed at every joist with galvanized 8d ring-shank nails. We ran the decking long and then cut it off after installation, leaving an overhang of 1½ in. beyond the aprons. We rounded the edges with a router, sanded the deck and finished it with three coats of the porch and floor enamel. We use oil-based enamel because it seems to dry harder and last longer than latex enamels, and it also has a high gloss we just can't get from latex paints.

Adding new skirt sections—The new skirting, which covers the gap between the porch deck and the ground, consists of frames made from ¾-in. lumber and either solid or lattice panels. Our skirts were designed to use minimal materials and allow maximum ventilation. And they maintain the original character of the Sheigs' porch. The top rail is the pine apron, 1x10 on the front and 1x12 on the sides (the side pieces are wider so that the top edges can be tapered to follow the pitch of the deck). These pieces were fastened to the deck framing with 2-in. narrow-crown galvanized staples. The stiles are 1x8 pine, and the bottom rail is 1x8 pressure-treated yellow pine. The stiles and the rails are joined with biscuits, and the joints are backed with wood blocks glued with construction adhesive and stapled from behind. The support framing for the skirting was kept to a minimum to avoid a foothold for decay. We used either horizontal 2x4 kickers attached to the posts or triangular braces made from 2x4s (drawings p. 18).

The skirting on the original deck was a mix of lattice and solid panels, and we matched what was there. The solid panels are ¼-in. lauan mahogany plywood with two coats of WEST System marine epoxy (Gougeon Brothers, Inc., P. O. Box 908, Bay City, Mich. 48707; 517-684-7286). The marine epoxy doesn't raise the grain on the thin plywood panels and gives them added protection from the elements. The panels were coated on both sides and on the edges. We mounted them on the inside of the framework with ¾-in. galvanized staples; an ogee panel molding finishes the perimeter. The lattice panels were built in place using ½-in. by 2¾-in. pine strips, which we ran vertically and horizontally.

Building the stairs—We cut our stringers from 2x12 pressure-treated yellow pine, which we buy in bulk and season for about a year before using it. We set the stringers 16 in. o. c. and attached the top ends directly to the outermost floor joist. To stabilize the stringers and to prevent twist, we installed 2x4 blocks between the tops of the stringers but held the blocks away from the face of the joist with ¼-in. furring (photo above). The bottoms of the stringers were set on a concrete pad pitched to carry water away from the porch. Once installed, the stringers were coated with two coats of the same preservative we used on the support posts.

We used clear pine for both risers and treads because pressure-treated lumber is more susceptible to cracking and warping than clear pine is and because it doesn't have the same finished look as clear pine when painted. But we made the first riser from pressure-treated lumber because it rests directly on the ground. Risers are made from ¾-in. stock that is ripped so that the top edge supports the front edge of the next tread. We leave a ⅜-in. space between the bottom edge of the riser and the stringer to eliminate a water trap. The treads are made from 5/4 stock, two for each step. We also leave a 3/16-in. gap between the two treads and between risers and the adjoining treads so that water will drain easily. To provide sure footing, we epoxied a strip of crushed walnut chips on the front of each step. The chips are manufactured by Buffalo Sand Blasting Sands Company, Inc. (25 Katherine St., Buffalo, N. Y. 14210; 716-852-4181). We use size 10-12 chips and find they work better than anything else we've tried.

A new main beam and roof—We had hoped to save most of the original hip-roof framing and the supporting beam. But we ran into a typical renovation dilemma—more rot than we had expected. We stripped five layers of leaky roofing and removed the 1x8 plank sheathing, which had rotted in spots, badly at the edges. The decorative rafter tails had deteriorated and so had the decorative T&G beaded-pine sheathing used on the overhanging portion of the roof. The front section of the beam had taken water from above and, because it was made up of sandwiched 2x8s and wrapped tightly in ¾-in. pine, the beam had trapped the water and rotted.

To get the strength we needed and still have a relatively hollow front beam, we used a 2x10 Micro-Lam (a laminated plywood beam), which ran the entire length of the porch. To build it out to the necessary finished dimensions, we glued and screwed 2x3 blocks to one side of the Micro-Lam and then glued ¼-in. vertical furring to the other side. We also furred the bottom of the beam with ¼-in. material, then wrapped the sides and the bottom in ¾-in. pine to match the existing side beams.

We drilled a 3-in. dia. hole in the bottom of the beam where it would rest on each of the 10 columns. The holes would allow air circulation between the hollow columns and the beam and roof framing. We cut new rafter tails from clear pine, sealed them with two coats of WEST System epoxy and installed them. New 3-in. T&G beaded pine was run on the top of the rafter tails where it could be seen from below. From this point up, where the sheathing would be hidden by the porch ceiling, we used ¾-in. CDX exterior plywood. We used 30-lb. felt and asphalt self-sealing three-tab shingles to finish the roof.

Vented columns and newels—To me, installing columns and newels is the most enjoy-

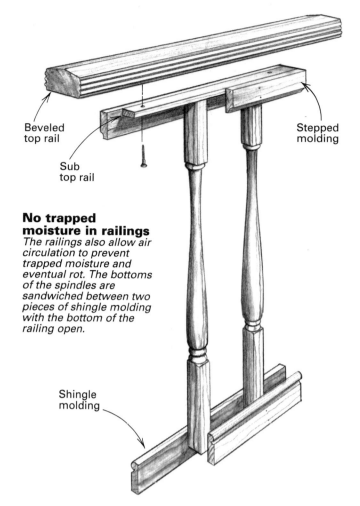

No trapped moisture in railings
The railings also allow air circulation to prevent trapped moisture and eventual rot. The bottoms of the spindles are sandwiched between two pieces of shingle molding with the bottom of the railing open.

Beveled top rail

Sub top rail

Stepped molding

Shingle molding

able part of the job. This is what brings neighbors out of their houses and what seems to invite people to stop their cars and visit. The 10, 8-ft. tall, 6-in. by 6-in. columns and the four 3-ft. high, 6-in. by 6-in. newel posts (drawing p. 44) were made from ¾-in. clear pine. The posts are simply long boxes primed on the inside and stapled together with galvanized 2-in. narrow-crown stables.

The trick to keeping air moving through both the columns and the newel posts is wooden plinths we make ourselves. Making the plinths takes time, but they add a finished look to our porches (compared to commercially available metal vented plinths), and we can make them any size we need. We start with a base made from shaped 5/4 by 2½-in. clear pine. We miter and spline the corners and glue the base together with epoxy, leaving a square hole in the middle for air to enter. We glue four 2½-in. square by ½-in. thick blocks to the bottom corners for feet. To the top of each plinth we then glue two blocks of wood that slide into the posts and prevent lateral movement. Once completed, each plinth gets two coats of marine epoxy. Although I don't do it, some builders install screening at the bottom of the plinth to keep out pesky insects.

The four plinths that anchor the newel posts have a 7/16-in. hole drilled through each corner. First we fill these holes with epoxy, and when it is dry we drill smaller holes through the epoxy for mounting screws. This effectively creates an epoxy gasket that prevents moisture from entering. We use galvanized 3-in. drywall screws to mount the newel plinths to the deck and the stairs. The plinths used under the columns are held in place by the weight of the porch roof and the beam.

Newel posts also need to be vented at the top, just like the columns. We start with a ¾-in. plywood subcap for each newel; we cut these to overhang the newels by 3/16 in. on each side. We then cut a 3-in. semicircular bite out of each side of the subcap. A 1-in. wide ogee molding hides the edges of the subcaps. A beveled cap made from ¾-in. clear pine is then fastened to each subcap, with a ⅝-in. overhang on all sides. Construction adhesive alone is used to install the finish caps to eliminate any nail holes and possible water infiltration. Once assembled, each cap gets two coats of marine epoxy inside and out.

When we are done, we have created a passageway for air to enter plinths at the bottom of columns and newels, travel upward and be vented at the top. This eliminates trapped moisture inside columns and newels and helps prevent rot.

New railings and finish—Once all the columns were built and installed, we took measurements for the 12 railing sections (drawing above left). For this porch, we used relatively standard railing components available from local suppliers. We started with turned spindles 1¾-in. square at top and bottom. They were sealed, sanded and primed before assembly. The spindles were spaced approximately 4 in. o. c. and held together at the top with ¾-in. by 1¾-in. subtop rails and on the side with stepped molding. On top of that assembly, we set sections of beveled top rail and sealed it with two coats of marine epoxy. To hold the bottoms together we sandwiched the spindles between two pieces of 2½-in. by 11/16-in. shingle molding. Water can't collect around the bottom of the spindles, and air can circulate freely. Railings are installed about 4 in. off the porch deck and are toenailed to the posts with 8d galvanized nails. The three long sections of railing were supported with 4-in. high pressure-treated blocks wedged between the deck and the bottom rails. We took care to seal the end grain on all components prior to installation.

We primed most of our components with an oil-based exterior primer before installation. This gave the parts some protection and stability during the four-week construction phase. After completion we filled the nail holes with glazing compound and brushed and sprayed on two coats of Pratt and Lambert Permalize Alkyd Gloss House and Trim enamel (P. O. Box 22, Buffalo, N. Y. 14240; 716-873-6000). Our only concern was painting the epoxied components, but paint adheres well to a sanded epoxy surface. □

Kevin M. Mahoney is a carpenter and supervisor with Victorian Restorations in Buffalo, N. Y., who also runs his own home-inspection business. Photos by the author.

Restoring a Porch

A fast, inexpensive method that doesn't require house jacks

by Roy F. Cole, Jr.

Some contractors use elaborate house jacks, and plenty of them, to repair two-story porches. Others jack and shore small sections, one at a time, eventually working from one end of the porch to the other. I've found both these methods too expensive because of the costly jacks and extra time involved. By restoring a fair number of two and three-story Charleston porches over the years, I've developed a method for repairing them well and repairing them fast. I shore up the whole porch with timbers, and never use jacks. They are just not necessary. I've done this on one-story porches too, but the challenge of the big ones is more to my liking.

Old Charleston is going through a major resurgence now, and there's plenty of restoration work going on. Luckily, the town was spared Victorian buildings because so few people had

The timber shoring on the porch at left is tied together to prevent workers from knocking it out of place. Cole's method of shoring allows for repair or complete replacement of columns and decking, as in the renovated porch below.

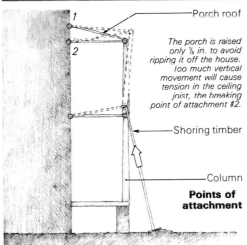

Porch roof

1

2

The porch is raised only ¼ in. to avoid ripping it off the house. Too much vertical movement will cause tension in the ceiling joist, the breaking point of attachment #2.

Shoring timber

Column

Points of attachment

A notch is cut at the upper end of the shoring (top photo) to receive the sill of the second-story deck. Cole places one 4-in. by 6-in. plank next to each column, at left. The base of the shoring sits on a 2-in. by 6-in. plank. Above center, an oak wedge driven between the shoring and the plank raises the second-floor deck. Once weight is off the column, a second wedge is driven under the first and nailed to the plank; then a wood block is nailed to the plank to stabilize the shoring.

any money to spend on houses from the end of the Civil War until the 1930s. So these porches, except for their poor structural condition, are originals. They are elaborate additions to wood or brick houses built after a major fire burned many of the wooden ones in 1861. They also survived an earthquake in the 1880s. Since porches were always popular in Charleston's hot, muggy climate, they were never removed, except for those that removed themselves.

To restore the two-story porch shown on the facing page, we had to replace column bases and plinth, some rotten decking, the sill and damaged hand rails. Luckily, the roof was in good shape, but it was a pretty extensive job nonetheless. We had to raise the second-floor porch deck and roof so the first-floor columns could be removed or lifted for repairs. The trick was to raise the roof no more than absolutely necessary. It was attached to the house at the two points shown in the drawing above, and over the years

had settled into a comfortable position. Disturbing this by too much jacking could have ripped the porch roof away from the house.

My method for taking the weight off the columns by shoring up the whole porch at one time allows us to perform each task only once. This saves a lot of time. Because of the weight involved we use 4-in. by 6-in. shoring timbers of southern yellow pine at each column, as shown directly above. I notch the tops of the timbers (top right photo) to receive the sill beam of the second-story deck. The angle of the notch is not critical. I place the base of the shoring timber on a 2-in. by 6-in. plank about 4 ft. long, and insert an oak wedge between the timber and the plank. The timber should be long enough so that its base is about 4 ft. from the porch; this leaves enough room to work.

Starting at one end of the porch, I tap the wedges with a heavy sledgehammer until each shoring timber relieves the weight on its col-

umn. Upward movement of no more than a fraction of an inch is enough to free the column from the upper-story sill. With the column free, I nail another oak wedge on top of the first and nail a wood block to the plank behind the wedge, as shown in the photo above. Friction between the 4-ft. plank under the shoring and the brick paving prevents the shoring from slipping backward. When no rough paving is available and the shoring must be supported on grass or bare ground, I increase the width and thickness of the planks, depending on the weight of the porch roof to be raised. For instance, I use 3-in. by 8-in. by 4-ft. planks (if the ground is fairly dry) to support shoring for a one-story porch. After all the columns are tapped free, I nail lightweight purlins horizontally to each shoring member to tie them together and prevent accidental movement. Then restoration work begins.

During porch restorations I do not remove columns or any other parts unless it's necessary for

Porch detail

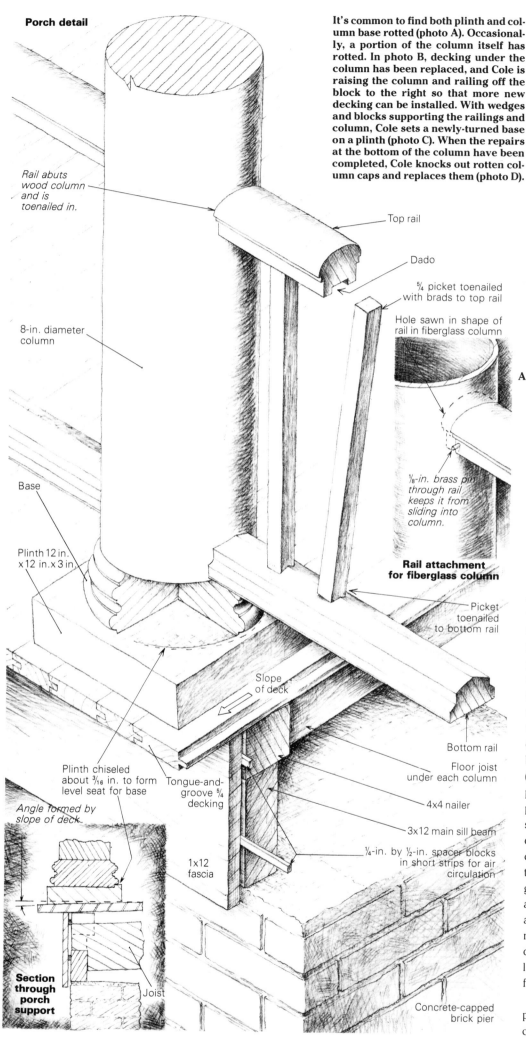

Rail abuts wood column and is toenailed in.

Top rail

Dado

⁵⁄₄ picket toenailed with brads to top rail

Hole sawn in shape of rail in fiberglass column

8-in. diameter column

Base

⅛-in. brass pin through rail keeps it from sliding into column.

Plinth 12 in. x 12 in. x 3 in.

Rail attachment for fiberglass column

Picket toenailed to bottom rail

Slope of deck

Plinth chiseled about ³⁄₁₆ in. to form level seat for base

Tongue-and-groove ⁵⁄₄ decking

Bottom rail

Floor joist under each column

4x4 nailer

3x12 main sill beam

Angle formed by slope of deck.

1x12 fascia

¼-in. by ½-in. spacer blocks in short strips for air circulation

Section through porch support

Joist

Concrete-capped brick pier

It's common to find both plinth and column base rotted (photo A). Occasionally, a portion of the column itself has rotted. In photo B, decking under the column has been replaced, and Cole is raising the column and railing off the block to the right so that more new decking can be installed. With wedges and blocks supporting the railings and column, Cole sets a newly-turned base on a plinth (photo C). When the repairs at the bottom of the column have been completed, Cole knocks out rotten column caps and replaces them (photo D).

A

repairs. Frequently the columns are original to the house; on the porch shown here, they are solid, heavy longleaf pine. When their bottom sections have rotted (as was the case with three columns on this job) they have to be removed. A 1-ft. length is cut off the bottom and the column is sent to a lumberyard to have that foot replaced and turned to match samples of the original. (This costs us about $100 per column.) Once the column has been scraped and repainted it is difficult to tell the old from the new.

There are times I have to completely replace columns. I use hollow, built-up wood columns, which, of course, are not as strong as solid ones that cost from $500 to $750 each. However, there is another replacement: fiberglass columns and plinth covers. When seen up close these columns do not look like wood, so I don't recommend using them in a true restoration. They are strong and rot resistant, but joining the wood rails to the column is a problem. To solve this, we carefully cut holes in the column for the top and bottom rails, slip them in, and pin them in place. (See drawing at left.) Also, because porch decks pitch away from the house for drainage, the plinth also slopes. The vertical column does not sit squarely upon it unless the plinth is beveled or chiseled to form a level seat, or the base of the column is cut on an angle. This is a critical detail to attend to when setting hollow wood and fiberglass columns. If the column does not seat evenly all around, stresses will build up, cause checking and eventually split open the column. The method I use is to put the plinth in place on the deck, then take a level and mark a horizontal line on the side of the plinth. I chisel out the seat for the base to approximate the angle drawn.

If the column is solid and only the base and plinth are rotten (photo A) which is the usual case, only the rotted parts are removed. To do

this, we place a block and wedge on each side of the column under the bottom rail. We hammer in the wedge and remove the base and plinth. When the decking under the plinth is also rotten, we remove it and replace it with new decking. Once the new decking is in place, we put a block and wedge directly under the column, transferring the weight to this block and the new decking (photo B). Then the decking on either side of the column is replaced as required.

Weight is placed back on the original blocks once rotten decking has been replaced, and a new base and plinth are slipped under the column (photo C). Then we remove the blocks on either side of the column. At this point, with no weight on the column, we knock out and replace broken caps (photo D). As we complete each column we loosen the shoring until the second story and roof are supported by the column.

If the porch substructure must be changed we do this before replacing the decking. The most common problem here is a rotten 4-in by 4-in. nailing strip behind the fascia where water gets in. I replace it with treated lumber and put a ½-in. spacer between it and the fascia to allow air circulation, as shown in the drawing at left.

Once all rotten columns, substructure and decking have been replaced and the shoring is removed, handrails are replaced. If rails are missing or destroyed I take samples to a lumberyard that reproduces millwork. New rails are cut to length, and their ends shaped to butt the round columns. They are toenailed in place with countersunk finishing nails. Then the old pickets are tilted and nailed into place as shown in the drawing at left. The porch is painted, and we move on to the next one. □

Roy Cole is a contractor who specializes in house restoration in Charleston, S.C.

Frugal Four-Square Fixup

Bumping out a wall and adding a porch
bring light and air to a turn-of-the-century classic

by Linda Mason Hunter

The new space. Though only bumped out 42 in., the entryway lets in light through a skylight and sidelights, opening up the dining room considerably. The pair of wide openings between dining room and porch links an indoor room with an outdoor room, and the house to the backyard.

When it was built in 1910, our house stood alone, surrounded by cornfields; it was a rural homestead on the Iowa prairie. Today it's only one of many houses on a tree-lined boulevard in Des Moines. The house is known as a "four-square" (photo, p. 29), and more specifically as a "homestead house." A four-square is simply a house that is predominantly square in plan (for more on four-square architecture, see the sidebar on p. 29). A homestead house is a type of four-square sold as a mail-order-kit house by Sears, Roebuck and Co., as well as by many lumberyards, from 1908 to 1913. The cost ranged from $733 to $853, including plans and all materials down to the nails.

Our house is really the quintessential farmhouse—a sturdy, unpretentious shelter that's simple in design, construction and decoration. Our remodeling aimed only to improve the house, not transform it.

Nice, but not perfect—The house was designed for 20th-century living, so it already fit our needs well. The kitchen, for example, wasn't a tiny afterthought designed exclusively for servants, as is often the case in houses built before 1900. Also, the house was originally plumbed, wired for electricity, and outfitted with central radiant heat (steam radiators), one of the simplest, healthiest means of heating. Finally, the floor plan is largely open, and it even included the original closets.

But our tidy house was not without problems. The interior was dark and the rooms were small. Our dining room couldn't comfortably fit more than six people at a time, making family celebrations cramped affairs. In the 1970s we made matters worse by stuffing an ancient air-conditioner into the only rear window in the dining room, making the room darker still and obscuring the backyard while eliminating any natural cross-ventilation. Also, though oriented well toward the front of the house, passage between the indoors and outdoors in back was inconvenient, making the backyard largely an unused space (top photo, next page).

Gradually, over a period of nearly a year, the design for a small addition evolved in my mind, a design that would remedy all of our house's shortcomings and create a more pleasant, livable space. I thought that by bumping the rear dining-room wall back 3 feet and making that space a light-well I would be able to brighten the entire first floor. To accomplish this I figured on fitting the new rear wall with French doors and sidelights, and adding a long horizontal skylight and sidewall windows. Finally, we would add a gazebo-like back porch, with access through the new French doors. Our new rear porch would mirror the front porch in style, materials, and detail. Though a relatively small project (photo above), it promised to open up the house to the backyard as well as provide my husband and me with a place to sit, read and talk.

An outdoor room realized. The back porch looks as if it had been built with the house. Most important, though, it provides a measure of privacy outdoors at home.

Paring and honing—I commissioned Bill Wagner, a long-time family friend and an experienced preservation architect, to work out the kinks in my design and come up with working drawings. Bill envisioned the bump-out as more of an entryway than an integral part of the dining room. To define the distinct functions of the spaces, he called for a pair of recessed-panel oak kneewalls to frame the passageway between dining room and entryway, the kneewalls mimicking those already separating living room from dining room.

Craftsman-style oak columns would be added at the inside ends of the new kneewalls to further reinforce the distinction between interior and transitional areas, as well as to add a little drama to the design (photo facing page). Across the top of the passageway an oak head jamb would tie the kneewalls and columns together visually while disguising the header.

Sitting around our oak dining-room table in February, Wagner, builder Bill Warner, and I worked out the details of the project. By substituting footings and piers for a foundation,

A dismal backyard. Before the porch and bump-out addition, the backyard wasn't much of a sight, and getting there required an awkard trek through the kitchen and pantry. This photo shows the layout of batter boards and stringlines.

A roof for the long haul. To support the roof sheathing, 2x10s were needed for strength. The exposed rafter tails were made thinner to give them a real delicate appearance.

Structure and sheathing. Many old porches have hollow columns supporting their roof, which is one reason you see so many saggy old porch roofs. To maintain the same look as the front porch yet give the back porch a considerably longer life, carpenters supported the roof with 4x4s notched and lag bolted to the joists. Scraps of 2x4 provide spacers and nailers for the red wood sheathing.

we'd be able to save a couple of thousand dollars at the outset. The new back porch would be built largely of redwood, as were the front porch and siding. The porch ceiling would be beadboard, painted sky blue ("to keep the flies away," according to an old homesteader myth). Interior trim would all be oak, finished to match the existing woodwork.

After carefully studying the plan, Warner submitted an estimate of $14,000, plumbing and electrical included. This was more than we'd wanted to spend, but we liked the design and details and felt we'd already eliminated everything we could. Because we wanted some landscaping, furnishings for the porch, and a financial cushion for whatever might go wrong, we secured a loan of $16,000—enough, we hoped, to cover all exigencies. With that in hand, we were ready to begin construction. Six weeks before we were scheduled to begin, we ordered the doors, sidelights, and skylight. Delivery can take a while, and we wanted these materials on site when they were needed.

Getting started—Like many remodeling projects, this one had its surprises. The first order of business was to have the electrical service and meter moved, as these were located on the wall to be razed. City building code, however, requires that whenever incoming power service and the meter are moved, all house wiring must be brought up to code. So we hired an electrician to identify and remedy any violations, and to move the service and meter.

Before we could begin laying out the porch or pouring the footings, we had to tear out the old back steps. In the process of removing them, Warner and his brother Craig discovered why our pantry leaned into the yard: a corner of the foundation was laid directly onto the ground—with no footings—and had settled over time. To halt the settling, Bill and Craig removed the problem foundation wall, dug a hole 4 ft. deep for a 10-in. square footing and rebuilt the wall on top of the footing.

After we attended to that unforeseen problem (and expense), it was time to begin the new construction. Bill and Craig laid out the addition with string and batter boards, then dug nine holes 4-ft. deep for the porch's 12-in. square footings.

Because the exterior bump-out wall wasn't going to be load-bearing, the new joists didn't have to be tied in to the existing floor joists. Instead, we lag bolted a ledger board to the existing band joist and attached the floor joists using joist hangers. To support the framing forming the porch's perimeter, we located 8x8 posts over each of the footings.

To support the porch roof, we used 4x4s, notched and lagged to the floor joists. Bill and Craig later wrapped each of these 4x4s with 1x8 redwood, using 2x4 blocks as spacers and nailers for the redwood sheathing (bottom right photo, facing page). Building porch-roof rafters and ceiling joists was relatively straightforward. The ends of the rafters were notched and ripped to 3½ in. where they would extend past the fascia to give them a lighter appearance that would

be more in keeping with the scale of the rest of the porch (bottom left photo, facing page). To keep the 20-in. overhang on the sides of the roof from sagging, 2x4s laid flatwise were laid into notches on the top of the two outermost rafters. Three of these outriggers, on 4-ft. centers, were used on each side to stiffen up the overhang framing. To give the end of the shed roof somewhat more character, Bill added a small gable.

Exterior detailing—To make brackets that would match those on the front porch, Bill and Craig traced the profile of one of the originals onto a piece of 1x12. Then, using the 1x12 as a pattern, they cut three 5/4x12-in. pieces of redwood for each bracket and laminated them together. The top and bottom pieces, which frame the brackets, are 4x4s with the ends beveled 45° on all four sides to form a point.

We duplicated the front-porch railing on the new porch. The top rail is 2x4 redwood with a 20° bevel on both sides of a 2-in. wide flat section that runs down the middle. The bottom rail is 2x4 redwood, laid on edge, with edges chamfered on both sides of a ¾-in. wide flat section. The slats are 1x4 redwood, 18 in. high, spaced an inch apart. These sit on the ¾-in. flat spot in the center of the bottom rail and are toenailed to it. Two pieces of cove molding, one on each side of the slats underneath the handrail, anchor the top of the slats to the handrail.

Bumping out—After stripping the exterior dining-room wall down to the studs, we were ready to cut the opening for the bump-out. Bill braced the ceiling temporarily with a 2x6 header on 2x4 "legs." With this support in place, he and Craig worked quickly to remove the old studs and frame in the new opening.

We built the two new kneewalls only 22 in. deep instead of 29 in. like the originals, so as not to block the view to the French doors. All other kneewall details and dimensions are the same as the original, including a bevel on the edge of the sill and on the top lip of the baseboard rail. We used cove molding over the panel, along the inside of the stiles and rails to match the original.

The core of each of the two columns is a 4-in. by 4-in. vertical box made of ¾-in. plywood. The box was toenailed to the oak sill; then, solid oak boards—the finish surface—were nailed to the plywood and to each other. Butt joints were used instead of miters to match the newel post in the front hall.

When the woodwork was done, it was my job to finish it, filling the holes with wood putty, sanding, staining, and varnishing. Having stripped, sanded, stained and finished all original woodwork 10 years before, I was able to get nearly an exact color and texture match with the new woodwork. □

Linda Mason Hunter is the author of The Healthy Home: An Attic-To-Basement Guide To Toxin-Free Living *(Pocket Books, 1990). Photos by author except where noted.*

Square architecture
by William J. Wagner

For practical reasons, square structures have long been popular. Perhaps the principal attraction of the square structure is that it possesses greater square footage than a rectangular (nonsquare) house with the same perimeter length. A 24-ft. by 40-ft. rectangle and a 32-ft. by 32-ft. square both have sides totaling 128 ft., but the square holds 1,024 sq. ft. as compared with the rectangle's 960 sq. ft.

Cubes and near-cubes are also very rigid structures. And because all angles are right angles, square houses are easy to lay out and economical to build. They're also very adaptable structures, permitting almost any kind of addition to the central plan without upsetting the design.

Doubtless, the square house's strength, economy and adaptability all account for some of its popularity as the four-square. With the advent of balloon framing the four-square blossomed in the U. S. Principal supporting members in balloon-frame houses are closely spaced 2x4 or 2x6 studs, which extend from floor to roof, and are used for both the exterior and key interior walls. The four-square lent itself beautifully to this new framing technique, with three rooms on the first floor (including the living room across the front half), a stair leading to four rooms on the second floor, and a third-story attic. Sills, joists, and studs were precut to size, then exterior walls and partitions were laid out on the deck and raised to a vertical position. Windows and doors were stacked above each other, so as not to waste any long studs in window or door areas.

The balloon-framed four-square was adapted to many different architectural styles in the early part of this century. From 1890 to 1920, the heyday of pattern-book architecture in the U. S., four-squares were the most popular style. These houses dot the countryside all over the Midwest and in the northern tier of states.

The most common manifestation of the four-square was as a hip-roof Prairie-style house (sometimes called an American Foursquare), although even Modern and International-style houses have adapted the practical four-square plan. Also, the basic four-square design would often take on secondary features of Mission-style and Italian Renaissance homes.

William J. Wagner has an interest in preservation, restoration and adaptive additions. He is restoration architect for many of Iowa's historical sites.

Two Lessons from a Porch Addition

A plumb-bob foundation layout and a cable-tensioned balustrade customize a tract house

by James C. C. Rice

New porch, new look. The original non-descript, split-level house (inset) was transformed with a new entryway (top). The new construction includes a roofed porch with a cedar-clapboard sunburst, an octagonal deck (at right) and a curving entry stair. The porch was built right over the old precast stairs.

Everyone's heard the story of the drunk who stumbles into a house down the street because it looks just like his own house. I guess the neighborhood sot could have shown up at Ed and Loretta Korzon's place. Their split-level ranch (inset, left) was built during the suburban speculative housing boom of the late 60s and was identical to the rest of the houses on the block.

In fact, the Korzons hadn't endured such an episode, and they might have had their front stairway to thank. Steep and treacherous, this precast stair pinned against the foundation was a struggle to climb—even while sober. The Korzons had pretty much given up using it. The basement shop door had become the primary entrance for the Korzons and their guests.

The Korzons wanted to use their front door again, so they hired me to design and build the covered porch with a curving front entry stair pictured here (photo above). Along with the gable-covered entry, I placed an octagonal deck in the northeastern corner of the porch. Its shape

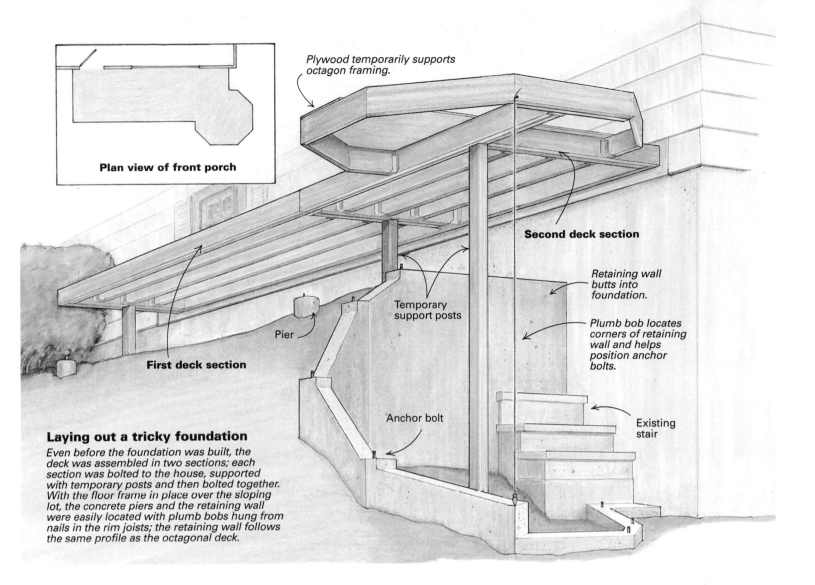

Plan view of front porch

Plywood temporarily supports octagon framing.

Second deck section

Retaining wall butts into foundation.

Plumb bob locates corners of retaining wall and helps position anchor bolts.

Temporary support posts

Pier

First deck section

Anchor bolt

Existing stair

Laying out a tricky foundation

Even before the foundation was built, the deck was assembled in two sections; each section was bolted to the house, supported with temporary posts and then bolted together. With the floor frame in place over the sloping lot, the concrete piers and the retaining wall were easily located with plumb bobs hung from nails in the rim joists; the retaining wall follows the same profile as the octagonal deck.

mimics a bay window on the house's north elevation. The deck offers a wonderful vantage point to view the distant hills and gives the house character. Laying out the octagonal foundation and devising a sturdy railing system for the octagonal porch were interesting challenges, and in this article I'll talk about how I dealt with both.

Frame first, foundation second—The Korzon house occupies a dynamic, steeply sloped corner lot. Finding a way to lay out the new porch foundation accurately along the steep, irregular ground was the first problem.

Half the porch is supported on 10-in. piers. For the other half, including the octagonal deck, I designed a sloping octagonal retaining wall that returns into the existing foundation (drawing above). Why a retaining wall? In the first place, I decided to build the new front porch over the old precast staircase. Only the top landing and the bottom two steps were removed. I wanted to use the material excavated from the pier holes to level out the ground below the porch and bury the precast stair. The retaining wall holds back this excavated material. Also, piers supporting each corner of the octagon would have been so close together that it was easier to dig a trench.

As I tried to figure out how to lay out the foundation, I realized how much easier it would be if the porch floor were already in place. Then I could just drop plumb bobs down from the framing to locate the piers and the retaining wall.

Plumbing the forms. Plumb lines at the deck corners help locate the foundation forms.

Beveled supports. Posts were milled to continue the octagon down to the retaining wall.

Building the floor frame first would simplify the task of laying out the foundation. So now my problem was figuring out how to install the porch floor without a foundation to support it.

Framing the floor—I could frame the porch floor most accurately by building it on the ground and raising it in two sections. To construct the section that includes the octagon, I built a template from two pieces of plywood

screwed to some 2x4s. On it, I drew the octagon's framing, trim, and column and newel-post locations at full scale. The template reduced the math necessary to determine the length and shape of the framing members. Often I just scribed individual members right from the template.

The octagon cantilevers over the porch's rim joists, so I tacked a sheet of plywood on top of the floor framing to hold the octagon in place (remember, I've got no foundation yet). Then

Cable-tensioned balustrade

Marking the cable hole. Railing sections were held between newels with a rope so that the newels could be marked and bored for the cable that strengthens the handrail.

A cable channel. The rough newel was cut to size, then a channel was routed for a length of copper tubing. The cable passes through the tubing to protect the wood from damage.

To cap it all off. Cut on a table saw from clear 2x8 cedar, the newel-post caps were sanded and varnished. A block screwed under the cap fits into the newel post.

Handrail conceals cable. This cedar handrail was grooved to hide the cable. Once the cable was tight, the handrail and a corresponding rail at the bottom of each section were fastened to the newels with galvanized 16d finish nails.

Anchoring the tension ring
The tension ring, a length of stainless-steel cable, is anchored at the columns that flank the octagon. The cable loops through the eyebolt and is clamped tight. Cable tension is adjusted with a socket wrench.

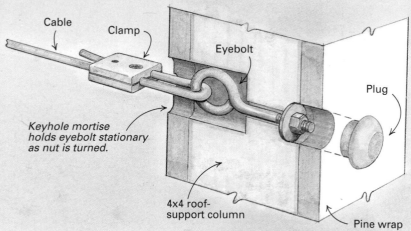

Cable
Clamp
Eyebolt
Plug
Keyhole mortise holds eyebolt stationary as nut is turned.
4x4 roof-support column
Pine wrap

---

each section of the floor frame was lag-bolted to the house and to each other.

Supporting the edge of the floor frame temporarily was easier than I had imagined. Three 4x4 columns, 2 ft. in from the edge, did the trick and left me room to dig the pier holes and the trenches and to place the concrete.

After bolting the two floor sections together, locating the piers was easy. I started a few nails along the rim joist and hung plumb bobs.

I did the same for the retaining wall, which bears on a concrete footing. I dropped plumb bobs from each corner of the deck framing, measured 1 ft. to both sides of the bob and dug a circular trench below the frost line. The ground was firm, so it became the form for the footing.

When it came time to pour the retaining wall itself, I built forms to match the profile of the octagonal deck (left photo, p. 31). The forms sat on the footing, plumb with the deck framing. To get the sloping effect in the retaining wall, I built hinged plywood doors at the top of

the forms. As the forms filled with concrete, I closed the doors, tacked them shut and continued the pour.

Finally, I used plumb bobs to locate the anchor bolts in each corner of the retaining wall and in the piers. Later I attached galvanized steel post anchors and installed permanent pressure-treated posts to support the porch and the deck.

Shedding water—The porch floor is 5/4 Douglas fir tongue-and-groove decking that runs perpendicular to the house. It's important to pitch exterior T&G decks because water cannot drain through the joints. In this case, I installed the posts along the rim joist 1 in. shorter than level to facilitate water runoff. The posts supporting the octagonal deck were cut to match the same slope, and their faces were milled at 22.5° (right photo, p. 31) so that the trim below the deck would attach smoothly to the rough framing.

The decking was back-primed prior to installation. I let the decking run long, trimmed it to size

and routed the edges with a roundover bit. After sanding the surface, I finished the deck with two coats of Benjamin Moore Moorwood Deck Stain (Benjamin Moore & Co., 51 Chestnut Ridge Road, Montvale, N. J. 07645; 201-573-9600) tinted light gray. This stain penetrates the grain and repels water like wax. It's a superior finish but requires annual reapplications.

Completing the porch—With the decking installed, I was out of the hole, and the rest of the framing was straightforward. The 5-in-12 gable porch roof was framed on top of the house roof, and the outboard end rests on a built-up carrying beam and four 4x4 cedar columns. The new roof follows the same fascia and overhang lines as the existing roof, but it has a rake overhang.

I decided early on not to put a roof over the octagon because it's the section of the porch that stands highest above the ground, and a roof over this high section would overwhelm the rest of the house. The railing around the octagonal deck works much better with the sloping site because it serves as a stepped transition between the roofed section and the ground.

However, I couldn't put a column at each end of the roof because the one at the northeastern corner would land in the middle of the octagonal deck. Instead, I put a column on both sides of the octagon, and the roof's carrying beams cantilever over these columns. To balance visually the pair of columns at the octagon, I placed two columns on the other end of the porch. Both pairs create thresholds—one frames the view (photo right), the other frames the front door.

After installing ⅜-in. Douglas fir ceiling bead and trimming the soffits and the carrying beams with clear cedar, I fabricated a sunburst at my shop and nailed it on the gable end.

Below the porch, I trimmed the rim joist, the posts and the lattice in clear cedar. The square lattice I used was hard to find, but it works much better with the porch design than common diagonal lattice would have. With the bulk of the trim out of the way, I turned to the project's final challenge: building the octagonal balustrade.

Cable strengthens railing—Because an octagon is strong in compression but weak in tension, octagonal railings tend to be loose and rickety. The railing I built for the octagonal deck has square-cut handrails that butt into five-sided newel posts. I worried that this railing would loosen when people leaned on it. Then a light went on: How about a tension ring?

A tension ring is commonly built into the top plate of a circular building with a conical roof. The tension ring keeps the walls from splaying under the roof load. In the octagonal railing I built, the posts and the top and bottom rails act in compression, and the tension ring—a continuous steel cable threaded through the posts and anchored at the columns—provides tensile strength. People can lean against the railing until they cramp up, yet it stays tight.

At the local marina I found plastic-coated stainless-steel cable and some flat-profile clamps; at the hardware store I bought two 4-in. stainless-steel eyebolts with nuts and washers, and at the plumbing supply store I purchased a short length of ⅜-in. copper tubing. I used the copper tubing as a sleeve to protect the wood fibers where the cable passes through the newel posts. I put the railing together and held it tight with a rope so that I could mark where the cable would enter the newel posts, which were not yet cut to length.

The newel posts are wrapped in clear pine; I made the wraps separately and slipped them over the rough newels. Next I set the railing sections (which I had preassembled) on 2x4 blocks between the newels. Then I scribed the posts (top left photo, facing page).

I removed the railing sections, then used a ⅜-in. wood-boring bit to plunge through the wraps into the rough posts about ¼ in. The drill marked precisely where the cable threads through the rough posts. I slid the wraps off and inserted a short length of copper tubing in each one. Next I cut the rough posts off directly above the ⅜-in. boring-bit marks. I then slid the wraps, with the copper tubing installed, back over the rough posts

View from the octagon. The octagon's height and location make it a good lookout point, and the cable-tensioned handrail adds security. Above, the corner of the porch roof cantilevers over columns that frame the view. The decking is 5/4 Douglas fir finished with a clear deck stain.

and traced the tubing's outline. I used a router with a ¼-in. straight bit to cut a channel for the tubing (top middle photo, facing page).

Anchoring and tightening the cable—Installation of the cable was a breeze. The cedar handrail stock I purchased has a milled groove on the underside (bottom photo, facing page). This groove was a fine place to conceal the cable. The cable runs beneath the handrail, through the center of the newel posts and ties to the two support columns flanking the octagon.

The cable is anchored at the support columns with eyebolts (drawing facing page). I scribed the entry point for each eyebolt shaft when the railing was held temporarily and bored a hole. Above and below this hole I drilled others to make a keyhole mortise. This mortise prevents the eyebolt from turning as it's tightened.

I slipped in the eyebolts, threaded the nuts, passed the cable through the eyes and clamped it. I had to widen the channel under the handrail a little so that the clamp would fit. At the back of each column I drilled a hole big enough for a deep socket; once the cable was tight, I popped in a finish plug. Then I used 16d galvanized finish nails to attach the top and bottom rails to the newel posts.

The finishing touch is a five-sided pyramidal cap on each newel post (top right photo, facing page). I made the caps out of clear cedar 2x8 stock, sanded them and finished them and the top and bottom rails with Benjamin Moore Impervo 440 Spar Varnish. ☐

James C. C. Rice designs and builds custom homes with Atlantic Contractors in St. Thomas, USVI. Photos by the author except where noted.

Photo this page: Rich Ziegner

Two-Story Summer Addition

A redwood porch and sun deck with sliding screen doors and decorative trellises

by Daryl E. Hansen

I share architect Frank Gehry's sentiment that many houses look better during construction than they do in their completed state. Anyone who has seen the exposed skeleton of a typical tract house and then compared it with the final product can appreciate this observation. In Minnesota—where thick, tightly sealed walls are the norm—this idea of exposed construction can be expressed best in summer spaces—porches and decks.

When it came time to remodel our 1911 house, the project included the addition of a screened porch with a sun deck above (photo facing page). In Minnesota, a sun deck in the right location can be usable even in February or March. Because I did all the work myself, the design was tailored to incorporate rudimentary woodworking skills, building on site with dimensional lumber and simple tools. The framework includes the outer trellis layer, which provides a building edge that orients toward the outside as well as shelters the habitable inner layers of porch and deck. This layered framework allowed the detailing to evolve well into the construction phase, with certain details appearing just as they were needed, such as the curved handles on the sliding screen doors.

Except for the porch ceiling, all the wood used in the addition was redwood. Clear redwood was selected for highly visible items, such as the trellis members, and for the remainder construction heart was used, which is a grade of redwood that contains knots of various sizes and other slight imperfections. The dimensions of the overall structure, the flooring, the screens and the trellis supports were designed to minimize cutting and waste of standard lumber sizes. Being both the architect and the builder allowed me to hand-pick the lumber and to purchase exact lengths and quantities so that lower-floor and upper-deck members have no joints.

Screened porch—To increase the spacious feeling of the lower-level screened porch, the ceiling is 9 ft. high with 8-ft. high screens on three sides (photo right). Flexible outside space was an important design criteria. Running the length of the porch on both sides are two pairs of screen doors, each with one fixed and one sliding panel. The screen doors slide open to incorporate the surrounding deck and trellised areas during those months when in-

sects aren't a problem. This effectively changes the size and feeling of the space, similar to the way *shoji* function in traditional Japanese house design. The demarcation between interior and exterior is blurred, and the garden becomes a source of decoration inside.

A vine trellis surrounds the screened porch on two sides. Eventually a dense covering of Engleman Ivy will create a "room around a room." This second enclosure will also provide privacy from neighbors—a necessity on a 40-ft. wide lot. Openings in the trellis will create "windows" in the ivy. The trellis also acts as a transition from building to garden.

Structural detailing almost inevitably becomes part of the design when the structure itself is exposed, and this project was no exception. For example, where a 4x6 post was required for a column supporting the ridge, I used three members instead—two 2x4s on either side of an elliptical center section cut from a 2x12. The overall shape and scale of this post (and others) is that of a stylized human figure. On the gable end of the structure, the center sections of these posts continue

from level to level, which helps the structure to resist wind forces. All other roof-trellis supports are similarly cantilevered.

Sliding screens—The sliding screens operate on noncorrosive sliding-door hardware (Grant Hardware Co., Div., Grant Industries Inc., High St., West Nyack, N. Y. 10994). Screwed to the top rail of the doors, the two top-mount carriers each have a pair of nylon rollers and are adjustable vertically. I chose top-mount carriers (as opposed to side mounts, which attach to the face of the rail) so that they would not be visible. The overhead track was installed in two 4 ft. sections to allow future repair or adjustment of the doors.

I used a single set of nylon guides, screwed to the floor where the fixed and sliding doors overlap, to align the bottoms of the doors. So far this has proven sufficient, but it would be possible to use a continuous floor guide that Grant makes. This guide would decrease the chance of the doors warping and would maintain consistent alignment along the bottom. On the other hand, a continuous floor guide

After building standard screen doors, Hansen cut two sets of curved-redwood trim pieces—one for inside, the other for outside—and simply screwed them together, sandwiching the screen between them. The screen doors along the sides of the porch slide open to incorporate the surrounding deck and trellised areas. Redwood louvers along the ceiling hide the light fixtures.

would be obtrusive visually and would either be obtrusive visually and would either be something to trip over or, if recessed into the floor, would collect water and dirt.

The short crescent-shaped pieces run vertically on each sliding door are attached to the decorative horizontal rails and serve as handles. These pieces are repeated at the top and on the outside of the fixed panels solely as decoration.

Simple detailing—The curved details of the sun deck and screened porch originated with the pair of screened French doors on the addition's gable end. After building the basic style-and-rail screen doors, I worked with cardboard patterns until I found the right radius for the curved details of the door. I cut two of each curved piece and simply dowelled and screwed the inside members to the frame. Then I secured the outside members with galvanized screws, sandwiching the screen between them. The horizontal rails in the middle of the doors cover joints in the curved pieces and strengthen the door.

The ceiling of the screened porch is made of beaded hem-fir boards. This material was commonly used in older porch ceilings and occurs in an existing screened porch on the front of the house.

At night the porch is illuminated by wooden cove lights running the length of the room (photo previous page). Controlled by a dimmer switch, the light is filtered through three wooden louvers. Spill slots direct some light

toward the floor, but most of the light reflects off the ceiling. I had the light fixtures custommade by Park Lighting, a local lighting firm. They are simple metal boxes, running the length of the room, with T-shaped double sockets projecting from them 12 in. o. c. The bulbs are linear-tube, 25-watt incandescents. A similar fixture, but with single sockets, can be purchased from Wiremold (The Wiremold Company, Electrical Division, 60 Woodlawn St., West Hartford, Conn. 06110).

I ripped the clear redwood louvers that form the cove to a thickness of ½ in., beveled their bottom edges and installed them on a slight angle to provide a unified flowing shape. The louvers are simply nailed to 2x supports cut in a stair-step shape. The horizontal lines of the wooden cove, accented by the hidden light, reinforce the detailing of the wooden screens and trellis members.

Sun deck—Anyone who has lived in Minnesota knows the value of a sunny day stolen from winter. Sheltered from the northwest wind, the upper-level sun deck is usable even at 40° F. I installed rafters over half of the deck so that the trellis on the north side of the structure could continue up over the sun deck (photo above). This creates the feeling of a room, but allows views to the sky and a sense of privacy from neighbors—even without vines. When morning glories eventually cover the trellis, shade will provide cover

to a portion of the deck. Openings in the roof structure, serving as skylights, will allow a view through the vines.

The railings and trellis members stop short of the house to avoid a collision of two separate textures. All four corners of the deck are recessed, or inverted. I did this on the side abutting the house in order to keep the railing from hitting the eaves. On the other end of the deck, the stepped-in corners create recesses for built-in benches. I made the bench under the trellised roof 34 in. deep and cut holes at the back for 13-in. clay flower pots. To provide lighting at night, I recessed a pair of lights—the kind used to illuminate outdoor stairs—in the front face of each bench.

The decorative trellis on the gable end of the sun deck extends the sense of enclosure and privacy in the outdoor room. The sunrise form frames the eastern view that inspired it. To lend a sense of scale to the 22-ft. long uprights cantilevered from below, an elliptical board, cut from a 2x12, was sandwiched between them. This served as a focal point for the radiating trellis. The bull-nose accent strips applied to the uprights (photo facing page) are details that appeared as they were needed.

A waterproof deck—Building an open deck over a dry porch created a waterproofing challenge. I started by building a standard

The trellis on the north side of the addition continues up the wall and across the rafters to create a sense of shelter on the sun deck (photo facing page). Once it's covered with vines, the trellis will also shade the benches beneath it. The author used simple joinery to striking effect. Connections were carefully nailed (photo above) so that the nail patterns would not detract from the orderly design.

Sun deck details

Beaded hem-fir

Cove light fixture

Redwood louvers

Top-mount carriers

Section through sliding screens

Strip of membrane glued to bottom of 2x2 protects continuous membrane.

Uncured flashing

¾-in. plywood

EPDM membrane

Metal flashing

EPDM counter flashing

1x4

Flower pot

Bench

1x2

2x4s continuous from first-floor joists to rafters.

floor-joist system with a ¾-in. plywood deck. Over that, I used a loose-laid roofing system called Sure-Seal (Carlisle SynTec Systems, Div. of Carlisle Corp., P. O. Box 7000, Carlisle, Pa. 17013). A single sheet of 45-mil EPDM membrane, large enough to hang slightly over the edges of the deck and flash up the wall of the house, was applied loosely and nailed around the perimeter (drawing right). Cutouts for posts were made before laying the membrane.

Next, an EPDM flashing (called Elastoform) was applied with mastic to the membrane at the base of posts. This is an uncured material that can be molded to particular shapes. I installed a pre-bent, prefinished metal-edge flashing along the perimeter of the roof, making sure that it covered the nailed portion of membrane and extended slightly over the edge. I then applied a layer of EPDM counterflashing to the EPDM membrane with mastic and lapped it over the metal-edge flashing to seal the joint.

The 2x6 deck boards are nailed to 2x2 sleepers in sections, four boards to a section. I glued a linear strip of roof membrane to the bottom of the sleepers to avoid any sharp contact with the roof membrane below. Carlisle offers a thicker rubber-mat material specifically for this purpose, but it was a special-order item and I didn't think it was necessary in a residential application. The deck panels are unfastened; their weight keeps them in

place and provides the necessary ballast for the loose-laid membrane system. The 2x2 sleepers of adjacent panels overlap so that the panels can be screwed together to avoid warping. Bench tops are also built in sections and screwed to a support frame. The ability to remove the deck boards and bench tops allows future repair to the roof membrane system.

Finishing redwood—I thought long and hard about the appropriate finish for the redwood. I wanted a clear natural-looking finish, which limited the selections. My research indicated that any natural finish would require yearly maintenance, but Sikkens (Sikkens Inc., 1696 Maxwell St., Troy, Mich. 48084) was advertising a three-to-five year maintenance cycle. Sikkens is a nonpenetrating oil/alkyd resin-based product. I decided to give it a try.

I added 20% of the mahogany finish to 80% of the natural (which Sikkens suggests in its literature) and applied this combination stain/varnish/water preservative in three coats. It produced a furniture-quality finish that really enhanced the wood grain. Unfortunately, the product has not held up well on horizontal surfaces where moisture and sunlight are constant. At the end of the first year, these areas had suffered fading and even some flaking from ultraviolet rays.

On protected vertical surfaces, the finish lasted longer, but it hasn't lasted three years, except in areas where little sunlight ever strikes. Subsequent maintenance of the finish sometimes requires two coats, but doesn't totally restore the look of the original wood. □

Daryl E. Hansen is an architect in Minneapolis, Minnesota.

Drawing: Michael Mandarano

Adding a Seasonal Porch

A timber-framed, screened-in porch with an awning roof keeps out insects but comes down easily to let in winter sun

by Ken Textor

Safe from bats and bugs. The seasonal porch keeps out afternoon sun in the summer and allows evening enjoyment of the deck. When cool weather arrives, the porch comes down for a sunny southern exposure.

I n an age of maintenance-free building, a structure that requires attention at least twice a year may seem out of step. But as I look back, my decision to convert an open deck into a summers-only screened-in porch still makes sense to me.

From the outset, you must believe that it actually gets hot in Maine during the summer. We live in a small clearing in the woods, sheltered by lots of tall, wind-shielding pines. On sunny days, the temperature on our open, south-facing deck regularly reaches well into the 90s, frequently topping 100°F or more.

Of course, by late afternoon it usually cools to the 60s. Unfortunately, the drop in temperature brings out mosquitoes that sometimes are difficult to distinguish from seagulls. In the evening, sorties of moths and formations of deerflies are chased by battalions of bats, ensuring that an open deck remains relatively unusable.

Then there's winter in Maine. It gets cold here. Very cold. To help with heating costs, you need all of the south-facing windows you can get. So shading those windows with a permanent roof over a screened-in porch that's only useful three months out of the year is counterproductive. Besides, such a roof would create a gloomy, sunless living room, which in my house is the south-facing room adjacent to the deck. So the idea of a seasonal screened-in porch was born (photo facing page).

Choosing the right type of beams was the critical first step—I couldn't find a design for such a porch in any architecture book, so I fell back on my own building knowledge, particularly my post-and-beam background and some of my boat-building experience.

The post-and-beam approach seemed the best way to achieve the required basic strength without a lot of studs to clutter the view. A minimum of framing members also would simplify assembly and disassembly. Because the load on these posts and beams would be no more than the weight of the galvanized steel-tubing awning frame and the awning itself, 4x4 top plates and posts would be more than sufficient.

Still, I had to keep everything lightweight for easy assembly and disassembly by one person. I also knew that posts would absorb rainwater along their bottom edges like sponges. So the post-and-beam structure would also have to be highly rot-resistant. These requirements forced me to rule out pressure-treated pine. Although available and inexpensive, it's too heavy and prone to movement during the humidity swings common during summers on the Maine coast.

The need for a light, durable wood led me to cedar. I could've used local northern white cedar, but I settled on western red cedar as the

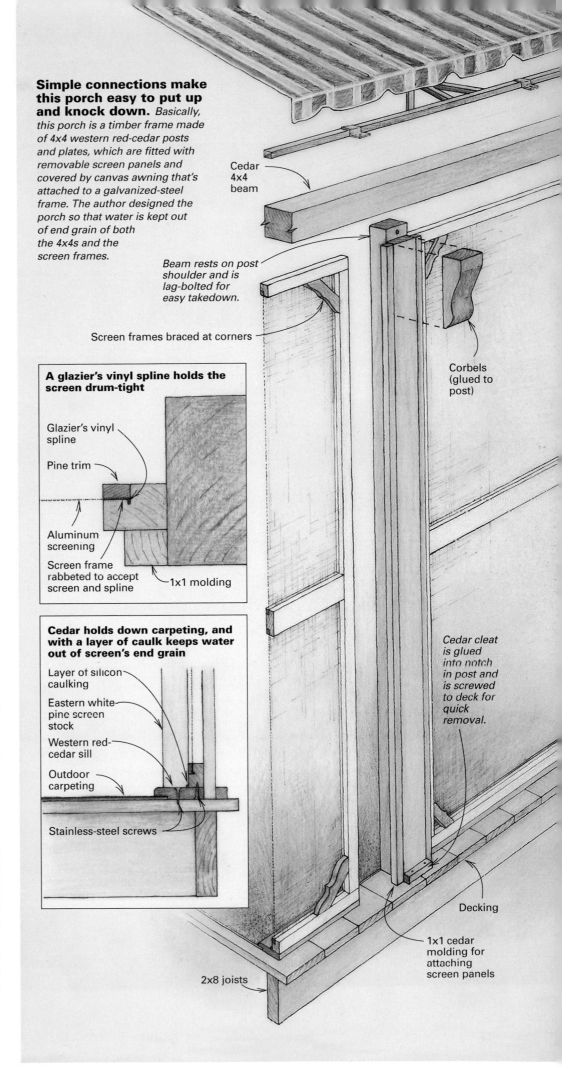

Simple connections make this porch easy to put up and knock down. *Basically, this porch is a timber frame made of 4x4 western red-cedar posts and plates, which are fitted with removable screen panels and covered by canvas awning that's attached to a galvanized-steel frame. The author designed the porch so that water is kept out of end grain of both the 4x4s and the screen frames.*

Cedar 4x4 beam

Beam rests on post shoulder and is lag-bolted for easy takedown.

Screen frames braced at corners

Corbels (glued to post)

A glazier's vinyl spline holds the screen drum-tight

Glazier's vinyl spline

Pine trim

Aluminum screening

Screen frame rabbeted to accept screen and spline

1x1 molding

Cedar holds down carpeting, and with a layer of caulk keeps water out of screen's end grain

Layer of silicon caulking

Eastern white pine screen stock

Western red-cedar sill

Outdoor carpeting

Stainless-steel screws

Cedar cleat is glued into notch in post and is screwed to deck for quick removal.

Decking

1x1 cedar molding for attaching screen panels

2x8 joists

Diagonals and center beam ensure strength. Although it doesn't have to carry a great deal of weight, the seasonal-porch structure does have to be sturdy enough to hold the awning down during high winds and to last through many knockdowns and installations. The author used diagonal beams to brace at outside corners of the top plate and a center beam for added stability.

It's sturdy and lightweight, but it's bulky. Although the two sections of awning frame for the author's porch weigh only about 50 lb. apiece, they are unwieldy because of their size. One person can handle the frame, but it works better with two people on the job. If the frame had been in three sections instead of two, another beam would have been necessary.

best solution. It's more straight-grained than local cedar, and clear 4x4s are readily available in lengths up to 20 ft. California redwood also was a possibility, but it's slightly more expensive than western red cedar and would have taken longer to get delivered.

Diverting water away from the existing deck—The deck I planned to screen in was 12 ft. by 16 ft. The decking was eastern spruce, which

made it a good surface for solidly attaching a temporary structure. Spruce holds screws well, an important attribute for the design I had in mind, but it's not very rot-resistant, a point that forced me to adjust the design to keep water from being trapped between the screens and posts and the decking.

To do this, I offset the top plate by 2½ in., creating a small overhang on 4x4 posts, enough to keep the daily summer trickles of dew and driz-

zly fog from coursing down the screens and posts. Although the posts were securely notched and bolted to the top plate, I added a corbel to help support the top plate and soften the generally boxy look of the frame (drawing, p. 39).

At the bottom of the posts, I needed a lip through which I could securely fasten the post to the deck. Rather than use some sort of angle iron or metal bracket, I notched out a space at the bottom of the post. Then I glued and screwed a piece of western red cedar in the notch. In addition to eliminating the potentially ugly angle bracket, the cedar lip also cut down on the amount of water-absorbing end grain surface at the post bottom.

For additional rigidity, I used diagonal beams at the outside corners of the top plates (top photo). I also used a bracing beam down the middle of the structure to stabilize the outer wall top plate and to make one more surface on which the awning frame would rest.

Boat building taught me to avoid iron hardware—As a former boat builder, I decided that all of the bolts, screws and hardware in this project should be nonferrous. From long experience, I knew that even galvanized-iron hardware breaks down and starts to stain badly after a few years of coastal weather. The local hardware store carried a variety of stainless-steel hardware. But if you live away from the coast, a marine catalog can supply what you need.

Boat-building experience also led me to choose a two-part resorcinol glue for all of the joints on the screens and for attaching the corbels to the tops of the posts. Epoxy is a fine waterproof glue. But with most epoxies, you run the risk of joints becoming glue-starved. This condition happens when the two pieces of wood being joined are mated so well that too much of the epoxy is squeezed out under modest clamping pressure, weakening the joint. Resorcinol, however, works best in tight joints.

Screen-making wasn't as simple as it seemed—Construction of the screens seemed like a pretty straightforward process of making frames and putting screens into them. Fortunately, I talked with a local glazier first and learned a little about screens before I made some common mistakes.

The size of the screen area is important. To end up with drum-tight screens, it's important to keep the openings as small as possible. The absolutely largest opening in a screened wall or door should be no bigger than 3½ ft. by 5 ft. Larger than that, it's hard to keep the screen surface from bulging the first time it's bumped.

Screen stock also needs to be strong, light and durable. So I chose eastern white pine, the traditional stock for screen lumber. Western red cedar seems too brittle, even if I could have found the 5/4 and 6/4 rough stock. (As a general rule, the minimum finished thickness for screen stock should be 1 in. for screened walls, and 1¼ in. for screened doors.)

In constructing the screened walls, I used 1½-in. wide stock around the edges, knowing the screen would gain additional rigidity when screwed into the top beam, posts and deck (photo right). Likewise, the decorative diagonals at the corners of the screened frames would lend rigidity to the post-and-beam structure. I increased the size of the rail, dividing the upper and lower screen to 3 in. to compensate for its general lack of support.

All of the screened-frame and screened-door joints were half laps, which increased the structure's overall strength. Although mortise and tenon is the traditional joint and although modern methods tempted me to use biscuit joints at the corners, the additional strength of the half-laps seemed worth the extra effort.

The rabbet for the screening itself seemed simple enough until the glaziers at Coastal Glass in Bath, Maine, straightened me out again. I was simply going to rout a rabbet ⅝ in. wide and 5/16 in. deep. But to make the screening tight in the opening, glaziers now use a vinyl spline inside the rabbet (top inset drawing, p. 39). With the spline in place and the rabbet trim nailed home, the screen has little give at the edges and tends to remain tight.

To make the additional rabbet for the glazier's vinyl spline, I used a circular saw with a carbon-tipped finish blade in it, set it to the proper depth, set the blade guard against the screen frame and cut the spline groove.

The final step in constructing the screened frames was the addition of a cedar securing lip at the bottom of each frame. This step served three purposes: It gave me a lip, or sill, to use to screw the screen to the deck; it provided a means to hold down the outdoor carpeting; and it kept the screen edge a little farther away from water that might pool during rainstorms. I screwed the cedar lip to the bottom edge of the screen after I set the edge in silicone caulk to keep the pine frame from absorbing water (bottom inset drawing, p. 39).

I ripped 1-in. by 1-in. strips of cedar and mounted them along the inside top and sides of the posts. The screen panels are then screwed into these strips. One final note on screens: I advise against installing the screening yourself. Even professional glaziers have trouble keeping screening tight while they are setting the vinyl spline and stapling everything home. It's definitely a two-man job even for the pros, particularly with large screened surfaces. We chose charcoal-finish aluminum screening over shiny aluminum, nylon or brass.

Big screens need extra bracing and support. Wide screen stock and corner brackets prevent bulging screens, which are further stiffened when the screens are screwed into place. The cedar sill at the bottom of the screen holds down the carpeting and keeps water from the frame.

Of bugs, rugs and awnings—If you build a seasonal screened-in porch, it's wise to work closely with the company that will make and install the awning and its frame. Blackfoot Awning & Canvas in nearby Auburn, Maine, made several trips to the job site to double-check measurements and to discuss exactly how I wanted the awning to fit.

Many custom-awning companies work largely on storefront overhangs. Galvanized-steel frames support these awnings, which are usually laced down. For a screened-in porch, however, loose lacings would allow bugs to sneak inside. Instead, Blackfoot Awning used strips of Velcro where the awning met the wooden top plate. This construction effectively kept out the bugs.

Keeping bugs out also meant that we had to buy indoor-outdoor carpeting to cover the deck. When you buy the carpeting, be sure to ask about shrinkage. Hot sun will make some synthetic carpeting shrink. If so, buy a piece a little bigger than you need and cut it to size only after it's been in the sun a few days.

Finally, be careful to specify that the awning frame be lightweight and easily managed by one person. The welded frame for my porch could have been divided into three sections instead of two, making it easier for me to handle during breakdown and setup. In two sections, each piece is probably no more than 50 lb. But because of their bulk, they're awkward for one

person to handle (bottom photo, facing page). If you add extra frame sections, use additional beams to support additional sections.

Knockdown, storage and costs—At the end of the summer season, I decided to knock down my screened-in porch, post-and-beam frame included. Although the frame could survive winter's worst weather, taking it down is so simple that there's really no point in leaving it up.

Knocking the structure down took about three hours. I had some help during the process. But I doubt doing it by myself would have taken much more than an extra half-hour. Two people could have it all down and stored in two hours. Other than the awning frames, no piece weighed more than 15 lb.

The cost of the project was $2,200, including the all-weather carpeting, the awning and the frame. (The awning cost $1,000; the wood cost $550; the carpet cost $150; and miscellaneous glue and hardware cost $150.) The time to build all of the wooden parts worked out to about 14 working days for one person. The lion's share of that time was spent on the screens. There are, however, a few custom wooden screen manufacturers still around. That, of course, would probably double your costs.   □

Ken Textor is an author and woodworker in Arrowsic, Maine. Photos by Steve Culpepper.

A place to kick back and relax. The screen porch the author built on his own house combines Victorian detailing with a builder's considered construction methods. In the photo below, a single pressure-treated step runs all the way around the outside of the porch as a sort of plinth.

A Builder's Screen Porch

From a hip-framed floor that slopes in three directions to a coffered ceiling, a veteran carpenter builds his porch his way

by Scott McBride

My grandfather lived alone in a little bungalow by the seashore. We got to know each other in his final years by spending long summer evenings out on the screen porch. We talked about the many things the old man had done in his life and some of the things a young man might do with his. Sometimes we didn't talk at all—just listened to the waves and the pinging of the June bugs off the screen, watched the lights, smelled the breeze.

A screen porch at night can have a magic all its own, balancing as it does on the cusp between interior and exterior space. A porch offers just enough protection from the elements to foster relaxation and reflection, without shutting out the sounds and the smells of the cosmos. This dual nature of screen porches can make them difficult to build with style because the usual rules of interior and exterior construction often overlap in their design.

When the time came to build a screen porch on my own house here in Virginia (photo above), I had the luxury of time—no anxious client, no deadline and no hourly wages to worry about. So I included lots of special details that I hope will spare my porch some of the

Bottom photo: Jefferson Kolle

problems I've seen in 20 years of remodeling other people's houses.

The foundation—I sited my screen porch two risers up from grade and three risers down from the adjacent kitchen. This made a smooth transition to the yard without requiring too much of a descent when carrying an armful of dinner plates from the kitchen. To anchor the structure visually, I ran a continuous step of pressure-treated lumber around the perimeter as a sort of plinth (bottom photo, facing page).

The step is supported by pressure-treated lookouts that cantilever off the poured-concrete foundation (top photo, right). I used pressure-treated 2x8s for the lookouts, inserted them into my formwork and actually poured the concrete around and over them. There isn't much concrete above the lookouts, so to key each lookout into the mix, I nailed a joist hanger on both sides. A week after the pour, the projecting lookouts were rock solid.

A hip-framed floor—Masonry is the obvious choice for the floor of a screen porch because water blowing through the screens won't affect it. Also, in hot weather the coolness of a masonry floor feels good on your bare feet. On the downside, masonry is, well, hard. It's also difficult to keep clean, it's gritty underfoot, and it retains moisture in damp weather.

Open decking is a good alternative to masonry, as long as it's screened underneath to keep the bugs out. Spaced, pressure-treated yellow pine will make a good, serviceable floor, and having a roof overhead will protect the floor from the harsh sun that is the nemesis of pressure-treated lumber. But open decking looks utilitarian at best, and my wife and I wanted something a bit more refined.

I decided to use untreated kiln-dried yellow-pine flooring, bordered by a coping of treated 2x8 (middle photo, right). I have repaired a lot of old porches, and I have noticed that it's the outer ends of the old floors that eventually decay while the wood stays sound just a foot or so in from the drip line of the eaves. By bordering my floor with a treated coping, the untreated yellow-pine flooring would be recessed further under cover. Also, the coping would allow me to lay the tongue-and-groove (T&G) floor at the end of the job because the structure above—the roof and its supporting columns—bears on the coping, not on the flooring. A temporary plywood floor endured weather and foot traffic during construction and allowed me easy access to run wires in the 1-ft. deep crawlspace.

To ensure positive drainage, and to avoid standing water on the T&G floor I had decided to use, I pitched the floor ¼ in. per ft. from its center in three directions. This meant that I'd have to frame the floor like a shallow hip roof (bottom photo, right). What became the ridge of the floor framing was supported by concrete piers.

I ran 1x strapping perpendicular to the joists and eventually laid the flooring over the strapping. In addition to promoting good air circulation under the flooring, the strapping served two other purposes: It allowed the flooring to run

Thinking ahead. Lookouts embedded in the concrete (and held securely by the addition of a joist hanger nailed to each side) provide rock-solid support for the first tread of the step that runs around the porch's perimeter.

Coping with weather. A coping of pressure-treated 2x8s supports the porch posts. Weep channels in the coping and an aluminum pan divert rainwater blown through the screens.

Get hip. This floor system, which is framed like a shallow hip roof, allows water to run off the porch floor. Strapped joists bring the finish floor flush with the 2x8 coping.

parallel to the slope so that most of the water would flow by the joints in the flooring rather than into them. The strapping also brings the top of the 1x flooring flush with the 2x coping. I could have used pressure-treated 1x for the coping, but because the roof and its supporting posts rest on the coping, I wanted it to be substantial.

The joint between the ends of the flooring and the inside edge of the coping gave me pause. I knew that wind-driven water was likely to seep in here and be sucked up by the end grain of the flooring, leading to decay. I thought about leaving the joint intentionally open, say ¼ in., but I knew that such a gap would collect dirt and be an avenue for critters. Instead, I back-cut the ends of the floorboards at a 45° angle and let them cantilever a couple of inches past the strapping for good air circulation underneath. Meanwhile, the long point of the mitered end butts tightly to the coping.

To collect any water that might seep through the joint, I formed aluminum pans that run underneath the coping and lip out over the floor framing (middle photo, p. 43). I cut weep channels in the underside of the coping with a dado head mounted on my radial-arm saw to let water out and air in. I have since heard that aluminum reacts with the copper in treated wood, so I probably should have used copper for the pans.

Hollow posts and beams—The roof of a screen porch is generally supported by posts and beams rather than by walls. Solid pressure-treated posts work well for support, but they won't accommodate wiring or light switches. Solid posts also are prone to shrinking, twisting and checking.

I made hollow posts of clear fir, joining them with resorcinol glue. Biscuits provided registration during glue up (middle drawing, right). I rabbeted the sides of the posts to receive both the frames for the screen panels and the solid panels below the screens. The bottom of each post was rabbeted to house cast-aluminum post pedestals. The pedestals keep the bottoms of the posts dry. They also allow air to circulate inside the posts to dry up any internal condensation. Rabbeting the pedestals into the posts makes them almost invisible and ensures that all rainwater is carried safely down past the joint between the pedestal and the post.

Because the 2x8 coping on which the pedestals bear is pitched (because of the hipped floor framing), I used a stationary belt sander to grind the feet of the pedestals to match.

Inland Virginia where I live doesn't get the wind of the Florida coast, but we get plenty of gales, and last year a tornado ripped the roof off a Wal-Mart in another part of the state. To provide uplift resistance for my porch roof, I bolted the tops and bottoms of the posts in place. Rather than relying on weak end grain to hold the bolts, I ran horizontal pairs of steel dowels through the posts, 3½ in. from the top and the bottom (top and bottom drawings, right). The dowels were hacksawed from ⅜-in. dia. spikes. At the bottom I passed a lag bolt vertically between the dowels and screwed it down into the floor framing until the head of the lag came to bear against the dowels (bottom drawing, right). At the top I used a

Porch posts: construction and attachment details

To prevent uplift from strong winds, the hollow posts are bolted at the bottom to the 2x8 coping, and at the top to the rough beam.

Top of post

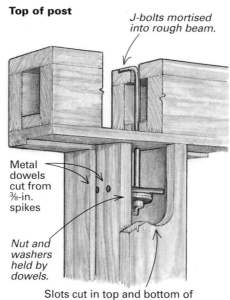

J-bolts mortised into rough beam.

Metal dowels cut from ⅜-in. spikes

Nut and washers held by dowels.

Slots cut in top and bottom of posts for wrench access

Middle of post

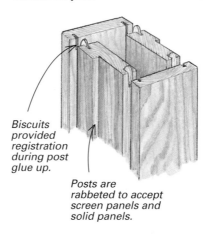

Biscuits provided registration during post glue up.

Posts are rabbeted to accept screen panels and solid panels.

Bottom of post

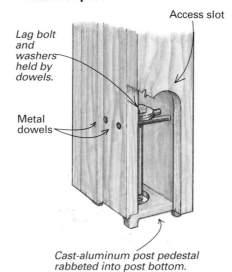

Access slot

Lag bolt and washers held by dowels.

Metal dowels

Cast-aluminum post pedestal rabbeted into post bottom.

similar arrangement, but instead of lag bolts, I used inverted J-bolts with the foot of the J mortised into the top of the rough beam, and the threaded end passing between the dowels. To get at the bolts with a wrench, I cut slots on the interior sides of the posts, which would be covered later with base and capital trim. I was surprised how rigid the posts felt after being bolted upright, even before they were tied together at the top.

The rough beams were made up with a box cross section rather than simply doubling up 2xs on edge (drawing p. 46). This gave the beam lateral as well as vertical strength so that any unresolved thrust loads from the untrussed secondary rafters above would be resisted by the horizontal top plate in the beam.

The roof and the ceiling—The inspiration for the coffered cathedral ceiling came from several sources. I once watched Japanese carpenters raise the frame of a small farmhouse. The delicate grid of the peeled white timbers against the sky made a lasting impression. I've also worked on Victorian houses in the Hudson Valley that featured finely wrought coffered ceilings over their verandas.

The framing scheme I finally decided upon is one that's found in some New England timber frames: trussed pairs of principle rafters interspersed with lighter, untrussed secondary rafters (middle right photo, facing page).

Instead of using heavy timber, I laminated each principle rafter in place from a 2x6 sandwiched between two 2x10s. Offsetting the bottom edge of the 2x6 helped disguise the joints, and the hollow channel above the 2x6 was useful for wiring.

Collar ties connecting principle rafter pairs have a 2x6 core sandwiched between 1x8s. The ¾-in. thickness of the 1x8 avoids an undesirable flush joint at the end where it butts into the rafter.

The secondary rafters are as wide as the principle rafters at the base, but their lower edges immediately arch up into a curve that reduces their width from 9 in. to 5 in. The constant width of all the rafters at the base allows the bird's mouth and frieze-block conditions to be uniform, even though the rafter width varies. I roughed out the curve of the secondary rafters with a jigsaw, then trimmed them with a flush-trim router bit guided by a template (top right photo, facing page).

Short 2x4 purlins span between the rafters on approximately 2-ft. centers (middle right photo, facing page). The ends of the purlins are housed in shallow pockets routed into the rafters, also with the help of a plywood template. I fastened the purlins with long galvanized screws.

The roof-framing material was selected from common yellow-pine framing lumber. Before I remilled the lumber, I stickered it and covered it with plywood for two months to let it dry.

The roof was sheathed with 2x6 T&G yellow pine run vertically, perpendicular to the purlins. The exposed V-joint side faces down, and the flush side faces up. Running the boards vertically added to the illusion of the porch's interior height; it was a pain in the neck to install because I had to maneuver from the eaves to the ridge while nailing each piece. To facilitate

Drawings: Bob Goodfellow

Yellow pine and Douglas fir complement one another on the interior of the porch. The rafter system, the vertical roof sheathing and the flooring are all yellow pine while the posts and the panels are Douglas fir.

Curved secondary rafters. To create the curves on the bottom edge of the secondary rafters, the author first rough cut the edges with a jigsaw, then trimmed them using a template and a router fitted with a flush-trim bit.

Primary and secondary rafters combined with a series of purlins comprise the porch's roof system. The secondary rafters curve along their bottom edges to reduce their width from 9 in. to 5 in. The purlins are let into the rafters and secured with screws.

Cluck, cluck, cluck. The author used a chicken ladder—a narrow set of stairs built on site—to ease the task of installing the vertical sheathing that runs from the eaves to the ridge.

Floor framing and post details

The porch is supported by a series of hollow posts. Plywood wainscot panels provide lateral rigidity. The wainscot panels and the shop-made screen panels fit into the rabbets cut into the posts.

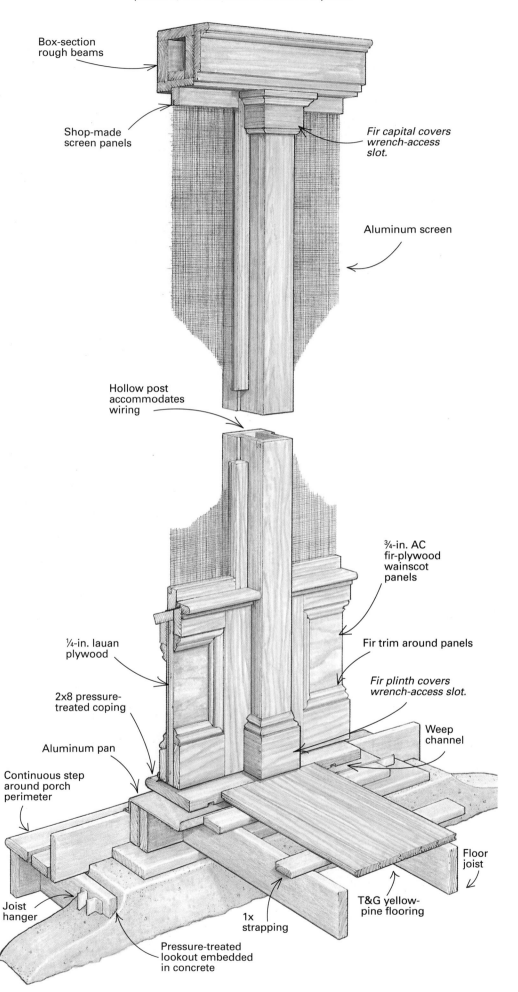

Box-section rough beams

Shop-made screen panels

Fir capital covers wrench-access slot.

Aluminum screen

Hollow post accommodates wiring

¾-in. AC fir-plywood wainscot panels

Fir trim around panels

Fir plinth covers wrench-access slot.

¼-in. lauan plywood

2x8 pressure-treated coping

Aluminum pan

Continuous step around porch perimeter

Weep channel

Joist hanger

1x strapping

Pressure-treated lookout embedded in concrete

T&G yellow-pine flooring

Floor joist

---

the process, I built a chicken ladder—a narrow staircase that hooks over the ridge and runs down to the eaves (bottom photo, p. 45).

Building a structure with an exposed finished frame was difficult and time-consuming. Floor space in my shop was strained to the max while all the components were fabricated. Everything had to be given multiple coats of a water-repellent finish to prepare it for the eventuality of rain before I could dry in the structure—I used Olympic WaterGuard (PPG Industries, Inc., One PPG Place, Pittsburgh, Pa. 15272; 412-434-3131). Moving ladders and scaffolding around all that finished woodwork was harrowing. The payoff, though, was a structure with a kind of bare-bones integrity that would have been hard to achieve with the conventional approach of rough framing wrapped with finish material.

Finish details—To contrast with the yellow pine in the ceiling and the floor, I used fir for all the woodwork from the floor up to the interior frieze (left photo, p. 45). The choice of fir allowed me to order matching stock screen doors, and this saved a lot of time in the shop. To reinforce the doors against racking, I introduced slender diagonal compression braces into the doors' lower screen panels.

The structure itself gains much-needed shear strength from the wainscot below each screen panel. The wainscot has no interior framing: It is built up with plywood and trim boards. First I screwed ¾-in. AC fir plywood panels to the posts, good side in. I bedded the panels into the same rabbets that would receive the screen frames above the wainscot. I then attached 5/4 fir rails and stiles to the inside face of the fir plywood. To avoid exposed nail heads, I screwed through the back of the panel to catch the trim.

On the outside, I tacked a sheet of ¼-in. lauan over the back of the AC plywood. Lauan holds up well in exterior applications and takes a good paint finish. The stiles and the rails on the outside were nailed through both layers of plywood into the interior stiles and rails. The resulting sandwich proved remarkably stiff. I capped the panels with a beveled sill and a rabbeted stool. For drainage, the bottom edge of the wainscot was raised 1 in. above the floor coping. To keep bugs out, I stapled a narrow skirt of insect screen around the outside. The top of this skirt was clamped down with a thin wooden band. A similar condition was achieved at the doors by attaching sweeps of insect screen. I even weatherstripped the edges of screen doors using a compressible-rubber weatherstripping (see *FHB* #78, pp. 92, 94).

When it came time to lay the T&G floor, I pondered the best way to deal with the shallow hips where the pitch of the floor changes direction. Rather than have a continuous 45° joint, which would be prone to opening up and collecting dirt, I decided to weave the floorboards in a herringbone pattern (top photo, facing page). Working from the longest boards out to the shortest, I grooved the end of each board so that it would engage the leading tongued edge of its neighbor. To cut the end groove, I used a ¼-in. wing cutter chucked in a router (bottom photo, right). The result is a pleasing stepped pattern that is accentuated by the way sunlight bounces off the wood according to the grain direction and the different planes of the hipped floor. Depending on where you stand, the floor has almost a faceted look; one side of the hip looks darker than the other.

Outside, I finished the porch with details consistent with my late 19th-century house. I extended the cornice return all the way across the gable by cantilevering lookouts off the gable studding. This creates a full pediment and gives the porch's gable end the same overhang protection as its eaves. The tops of the posts sport scroll brackets on the outside and simple capitals on the inside.

Screen for the porch—I made wood frames for my porch screens out of 1x2 fir. I used mortise-and-tenon joinery with an offset shoulder on the rails. The strength of a mortise-and-tenon joint isn't really necessary for a fixed frame that gets fully supported in a larger structure. But the design of a mortise-and-tenon joint makes it easy to use a table saw to cut the rabbets and plow the spline grooves before assembling the frame.

Spline stock holds the screen in the frame. Tubular in cross section, the spline stock gets pushed into a groove on the frame where its compression holds the screen in place. Spline stock is made from rubber or vinyl, and it's available in a smooth profile or with ridges around the circumference. The ridges help guide the splining tool, and they give the spline a little more bite on the walls of the groove.

The tool used to press in the spline looks like a double-ended pizza cutter. One disk has a convex edge used initially to crease the screen into the groove. The other disk has a concave edge, which tracks on the round spline as it is pressed into the groove.

The two most common types of screen are aluminum and vinyl. Aluminum screen is available in mill finish or charcoal (see *FHB* #54, p. 4, for a source of screen made of copper, stainless-steel, bronze, etc.).

I used mill-finish aluminum for my screen porch because it seemed to be the most transparent. I also think aluminum is somewhat stronger than vinyl and less likely to sag over wide spans. The main drawback of aluminum is oxidation, which gradually forms a grainy deposit on the wire and reduces the screen's transparency. I live in a rural inland area where salt and pollution aren't prevalent. If I lived near the sea or in an urban environment, I would have leaned toward vinyl. I would also go with vinyl if I were hanging the screen in place vertically, rather than rolling it out on a bench. Vinyl is much easier to work with and less likely to crease. A final consideration in choosing screen is the resounding ping made by bugs slamming into a tightly stretched aluminum screen. I rather enjoy it—it's one of the unique sounds of summer—but others might prefer to muffle the impact by using the softer vinyl screen. □

Scott McBride is a contributing editor of Fine Homebuilding. *He lives in Sperryville, Va. Photos by the author except where noted.*

The hipped floor slopes in three directions to shed water that blows through the screens (above). Sun hitting the finished floor gives a pleasing effect. The joists are cross-strapped, and the flooring is laid on the strapping so that it runs parallel to the slope of the porch floor. A router grooved the end of each piece of flooring (below) so that it could herringbone its way down the floor's hips.

Top photo: Jefferson Kolle

A Screen Porch Dresses up a Ranch

Wide overhangs and a wraparound stone patio keep this elegant outdoor-living area cool and dry

Scaffold provides work station for assembling trusses on site. Heavy 6x8 beams spanning from the house to the porch's corner posts carry trusses constructed of 2x4s and custom-made steel plates. Heavy beams allow for open walls that are sheltered by 2-ft. deep eave overhangs.

by Alex L. Varga

"Something there is that doesn't love a wall; that wants it down." I know this line from Robert Frost's *Mending Wall* evokes a deeper meaning; however, it seems that this sentiment applies whenever man endeavors to build a sound and lasting structure. There is always some element of nature or turn of fashion that conspires to alter the structure once it is completed.

I wonder what Frost would have written had he been building a wooden structure. I know I had my work cut out for me when I added a screen porch to a ranch house in Connecticut (top photo, facing page). My clients requested a shaded outdoor-living area that provided views and took advantage of cool breezes coming from the wooded area behind their house. The catch was that the outdoor-living area should have a finished wood floor and lots of wood trim. Digging the foundation 42 in. below grade to get under the frost line was only part of the solution. Here I'll discuss how I designed and built a porch to survive the elements as it provides a beautiful refuge from summer heat and pesky bugs.

Cypress frame over crushed stone—When planning the porch, I followed basic moisture-control strategies: allowing for drainage, creating ventilation and keeping wood out of direct contact with the ground. During the beginning stages of the project, work was required on the house's septic system. With the backhoe on site, I took the opportunity to cut the grade down about 2 ft. in the porch area. I also sloped the grade away from the house.

With the backhoe I then dug six 42-in. deep holes for the concrete foundation piers. Once the piers were in place, I bolted triple 2x10 beams around the perimeter of the porch, supported by the piers and bolted securely to them.

Taking the grade down 2 ft. allowed me to add a layer of crushed stone about 3 in. deep and still have almost a foot of clearance beneath the porch's floor joists. The crushed stone creates a clean and well-drained area under the floor.

The floor joists are 2x8 cypress, which costs a bit more than pressure-treated pine but is naturally, as opposed to chemically, rot resistant. Spaced on 2-ft. centers, the joists are connected with joist hangers to the perimeter beams and to a triple 2x10 center-span beam.

To prevent small animals from getting under the floor system, I attached galvanized-steel wire mesh to the inside surface of the perimeter beams, bent the wire back about 6 in. along the ground and covered the wire with crushed stone. During this phase of construction, I also prepared the site for a stone patio. The patio encircles the porch and extends slightly beneath it; the floor framing is about 3 in. above the perimeter stonework. Unlike the soil and plants that often surround porches, the stone patio allows rainwater to drain away from the porch, giving it a better shot at staying dry and ventilated. The patio also looks great and helps unify the new construction and the existing house.

Another basic moisture-control measure I took was to provide air circulation underneath the porch. I faced the perimeter beams with a cedar skirtboard that's ¾ in. above the stone. This ¾-in. gap runs continuously around the porch and lets air flow freely through the floor system.

Continuous headers and braced posts make for open walls—After putting single 2x4 pilasters at the house and double 2x4 posts at the corners of the floor frame, I installed two 24-ft. long 6x8 header beams on top of the pilasters and posts. The header beams run continuously from the house out to their free ends, which cantilever 2 ft. 6 in. beyond the freestanding posts that rest on the stone patio.

I wanted to keep the sidewalls of the porch open to catch summer breezes and to capture the view of the woods as clearly as I could. So I

The porch is part of a composition. The porch includes a stone patio and an octagonal, covered entry. These elements don't just look good, they help keep water away from the porch. In addition, the porch's roof features large overhangs and gutters for protection from sun and rain. Posts tapered to suggest the shapes of tree trunks support the gable overhang. Lights in the soffits shine on the roof and down around the porch perimeter.

Large screen openings celebrate fall colors. Tinted marine varnish on the fir flooring and decay-resistant cedar and cypress trim, primed on all sides, allow for interior-quality details in the screen porch. The trusses, made of small-dimension lumber, create a branchlike effect. In the soffit bays, 1x6 trim boards hide spotlights.

branches better than a framework of rafters and collar ties.

I fabricated the trusses on site using 2x4 fir studs and ⅛-in. galvanized-steel plates cut with a jigsaw. After making patterns, I cut all the truss pieces on the ground and assembled the trusses in place, working from a scaffold (photo p. 48). The trusses are on 5-ft. centers.

Between the trusses, I installed doubled 2x4 rafters. The rafters are braced at the ridge and at midspan by 2x4 kickers nailed to the trusses. Fanning out from the trusses, the kickers provided the branchlike effect that I had been pursuing.

Installed with the bevel face down, the 2x6 tongue-and-groove roof boards provide a little texture, and the dark lines of the bevel joints look good with the 2x4 framework.

Overhangs keep porch cool and dry—The porch roof overhangs the walls all around for shade and for rain protection. At the eaves, 2-ft. deep overhangs match the overhang depth on the existing house. The truss design allowed me to build these overhangs without using large collar ties. Such large members would have appeared much too heavy for the look I wanted. The trusses transfer all of the roof weight to the 6x8 headers, leaving the truss ends free to create the overhangs.

The large soffit overhangs also turned out to be a good place to locate floodlights to light the interior-ceiling surfaces. I hid these fixtures by adding a 1x6 trim detail to the inside top edge of the sidewall beams. Finished with exterior-grade fir plywood and painted to match the main house, the porch soffits are open to the interior. The gable end is also open, eliminating the need for any special soffit or ridge ventilation.

Curved columns resemble tree trunks—At the gable end, an 8-ft. overhang provides a covered section for sitting on the patio in addition to shading and keeping windblown rain from getting into the porch. This deep gable overhang rests on two curved columns I made in my shop. The columns are double 2x4s cased with primed and painted #2 cedar. I got the curved look by gluing and nailing wedge-shaped blocks of 1x stock to the tops of the 2x4 cores. I ripped both edges of cedar casing so that it was tapered, then glued and nailed the casing to the cores. The bottom of a finished column is a 5-in. square; the top is an 8-in. square. The casing conforms to the posts, giving the columns a curving profile similar to that of a tree trunk.

Spacing the roof boards creates a skylight—As the roof structure was nearing completion, I realized that although the porch would be comfortably shady during the heat of summer, it might end up a bit too dark. So before installing the last several feet of 2x6 roof boards, I began experimenting with ideas for a ridge skylight. Rather than cut an opening or a series of openings into the roof deck, I decided to take a hint from the lines of the 2x6 roof boards. In the top 2 ft. on both sides of the peak, I separated the boards after ripping off the tongues on a table saw. Starting with a ¼-in. gap, I widened each

He didn't run out of roofing boards. Because the roof has such deep overhangs, a ridge skylight was needed to brighten the porch. This unconventional skylight is a series of progressively wider gaps between roofing boards, all covered with a Lexan ridge cap. The trusses and 2x6 roof boards are finished with semitransparent stain similar to the gray of the floor and the stonework.

used single 2x4 posts on 5-ft. centers, which would later be fully cased in 1x trim, to frame the screen openings. Single 2x4 blocks at chair-rail height stabilize the posts against bending and twisting and make the finished screen sizes large but manageable.

The continuous 6x8 headers solidly connect the individual posts together at their tops and create a stable and rigid frame out of a row of rather slender posts. The casing of the posts in 1x trim and the horizontal bracing provided by the chair-rail blocks allow the single 2x4 posts to bear the necessary roof loads easily. To add to the rigidity of these sidewall frames, I used sections of 3-in. steel angle with 8d nails to connect the corner posts to the floor system and to the header beams. These angle brackets were later hidden underneath the finish trim work.

I could have used a bottom plate, but instead I opted to install the 2x4 posts directly on the perimeter beams (drawing facing page). I laid the flooring on the beams and capped the end grain with a 5/4x8 sill that's notched around the 2x4 posts. A double band of blocking topped with a 5/4x6 sill creates a wiring chase that, when covered with 1x4 trim, provides an attractive baseboard detail.

The site-built trusses create a branchlike effect—Inside the porch, I wanted the roof structure to appear light and open, almost treelike in its framework. My idea was to link the structure to the lacework of woods that the porch overlooks. So I designed a cathedral ceiling that would be supported by trusses built with small-dimension lumber, which evoked the image of tree

joint between boards to 2 in. at the peak (photo facing page).

Over these spaced roof boards I installed ⅜-in. thick double-wall Lexan sheets, which might be described as plastic, see-through cardboard. Lexan is obtainable from commercial-plastics distributors. First I folded each 4x8 sheet in half along its length, which was similar to folding a sheet of cardboard. Then I set the sheets in beads of silicone on top of the roof framing, leaving a ½-in. gap between the sheets for expansion. I then filled the gap with silicone.

I fastened the Lexan to the roof structure with battens made of ⅛-in. by 2-in. flat aluminum bar stock. I bent the aluminum to fit over the ridge, set the aluminum in silicone over the Lexan and drilled pilot holes through the aluminum and the Lexan into the roof framing. Then I screwed the bar stock to the frame, locking the Lexan skylight in place. This simple but effective skylight allows enough light into the porch to brighten it without overheating it. The resulting light from the spaced 2x6 roof boards is similar to sunlight filtering through tree branches.

Trimmed wall posts hold site-built screens—

For the trim on this project, I used a mix of cypress, red cedar and white cedar that I primed on all sides with an oil-based primer. Cedar and cypress have natural resistances against rot and decay. All of the trim was finished with an acrylic latex paint to match the existing house finish.

With the framing trimmed out, I was left with a pattern of 5-ft. square screen openings over smaller 5-ft. by 2-ft. screen openings. I made all the screens using aluminum frame stock and aluminum screen. I made most of the screens on the ground and popped them into their openings. Then I held them in place with 1x2 stock set with finishing nails. If a screen becomes damaged, the 1x2 is removed, and the screen comes right out.

In the upper section of the gable end, I had to fit screens into the triangular openings of a truss. I couldn't use manufactured square corner clips to join triangular frames, and the screen, which has a square cross mesh, wrinkled when I stretched it diagonally. I ended up screwing the frames in their openings and stretching the screen in place.

The southwest side of the porch is relatively exposed to both weather and neighbors. Here, I had a roll-down canvas storm shade installed to keep out windblown rain as well as the harsh southeast light in summer. The shade also is a privacy screen, blocking the porch's view from the neighbors' yard.

Tinted marine varnish protects strip flooring—

For the floor, I chose ¾-in. by 3½-in. tongue-and-groove fir to achieve a smooth, attractive and easy-maintenance surface. The added advantage of T&G boards for the floor surface is that they create an insect barrier, eliminating the need for screening around the skirt or under the joists. There's no plywood. The flooring runs directly over the joists.

The flooring's color, however, was overly reddish next to the bluish gray of the granite patio.

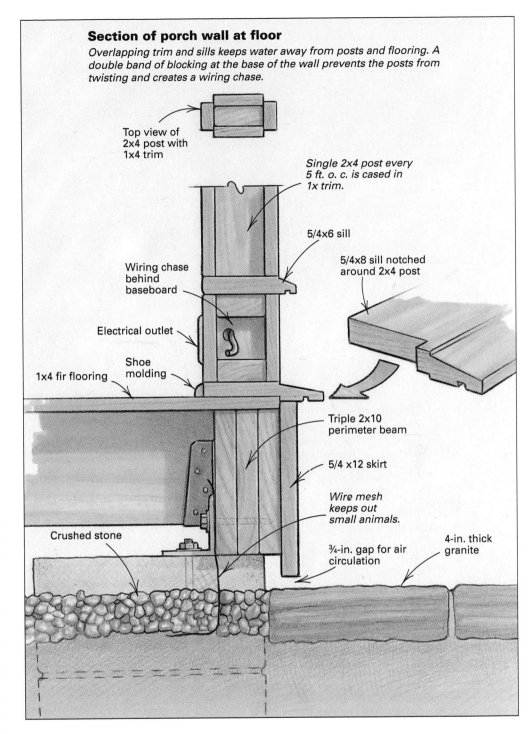

Section of porch wall at floor

Overlapping trim and sills keeps water away from posts and flooring. A double band of blocking at the base of the wall prevents the posts from twisting and creates a wiring chase.

Top view of 2x4 post with 1x4 trim

Single 2x4 post every 5 ft. o. c. is cased in 1x trim.

5/4x6 sill

5/4x8 sill notched around 2x4 post

Wiring chase behind baseboard

Electrical outlet

Shoe molding

1x4 fir flooring

Triple 2x10 perimeter beam

5/4 x12 skirt

Wire mesh keeps out small animals.

Crushed stone

¾-in. gap for air circulation

4-in. thick granite

So I decided to mix a tinted but still transparent high-gloss finish. After some research and experimentation, I chose a marine-grade spar varnish as the primary sealing element because it's waterproof and UV-resistant. I mixed the varnish with a small amount of bluish gray stain to help deaden the reddish tone of the fir. I applied four coats to the floor, resulting in a high-gloss, semi-transparent finish with an attractive grayish tint. The only drawback I discovered to mixing the stain into the varnish was that it seemed to slow the curing process. The varnish remained fragile, even though it was dry to the touch, for about two weeks.

With open-air porches it is common practice to pitch the floor approximately ⅛ in. per ft. away from the main building to allow standing water to drain. With this porch, however, I chose not to pitch the floor so that I could keep all of my trim

lines and screen openings level and parallel. I thought that the flooring's marine finish, the large roof overhangs with 4-in. aluminum gutters and the roll-down canvas storm screen would provide adequate protection to the interior from splashing water. This combination has proved itself effective.

As a second line of defense against water collecting on the porch floor, I installed ⅜-in. brass bushings on 5-ft. centers around the perimeter of the floor to serve as inconspicuous floor drains. These bushings function reasonably well; in the future, however, I probably will use a ¾-in. bushing to allow any intruding water to drain more easily. □

Alex L. Varga is an architectural designer in Hamden, Conn. Photos by Rich Ziegner except where noted.

Deck Design

A guide to the basics of deck construction

by Scott Grove

A properly built deck should last a lifetime. But for this to be possible, you must constantly think about nature's elements as you design and build it. If you don't, trapped moisture can promote bacterial degrade that will slowly eat your deck away.

Planning the deck—A deck is an intermediate space between the controlled environment of a house and the raw elements outdoors. Since a deck can expand the living area of the house and serve as an entry, it's important to consider traffic patterns in your planning. Avoid paths that cross through activity areas, and arrange for them to be as direct as possible. A path improperly located can isolate small areas and render them nearly useless.

A deck can accentuate the good features of an area and minimize the bad ones. It can conceal a fuel tank or snuggle around a tree. Decks are great for hiding ugly foundations, service meters or old concrete patios. Let these existing elements influence your deck design and they'll make your job easier. The space under a deck can be used to store firewood, too.

Safety is an important consideration when de-signing a deck. For instance, a landing in front of a door needs plenty of room to allow the door to open with at least one person on the landing. A low walkway that may be just fine without a railing in the summer can be most dangerous in the winter, when snow conceals its edges. Define these edges and all corners, using posts, trees, bushes, rocks or any other visual device on or off the deck that will help make the feature more obvious.

Designing a deck can seem complex if you've never built one before, so beginners should make a detailed drawing of the entire layout, board by board. Once the design is on paper, it's fairly easy to compile a list of materials. Planning the layout and orientation of a deck is at least as important as building it.

Estimating costs—We've been building decks in New York State for seven years, and we use the following figures for rough estimates of the materials and labor needed to build a deck: $8 per sq. ft. for decking (including the framing and footings), $7 per sq. ft. for stairs, $10 per lin. ft. for simple railings, $15 per lin. ft. for bevel-cut railings and $20 per lin. ft. for benches. These figures reflect our company's wage scale, construction speed and craftsmanship, and if a deck design is particularly unusual we'll adjust the figures upward. For those who work alone or with minimal help, the following materials-only estimate, based on prices for #1 pressure-treated lumber in New York State, will help determine approximate costs: $3.00 per sq. ft. for the decking lumber, $4.50 per sq. ft. for stairs, $2.25 per lin. ft. for railings, and $8.25 per lin. ft. for benches. The type of construction you use, and the level of detail you include, will have a significant effect on the expense of your deck.

Choosing lumber—Water is the worst enemy of woodwork, and this fact should be foremost in your mind as you select lumber for your deck. Remember that water does the most damage when it rests undisturbed on or in the wood, especially in places that are slow to dry out. Warping is the number-one problem with decks, and water contributes to the problem. Checks channel water inside a board to accelerate the decay process, and so when we're building a deck we routinely cut back boards with serious end checks. We allow for these cuts when we design a deck by making sure that our plans call for material about 6 in. shorter than standard lumber lengths.

The amount of moisture within new lumber determines how much it will shrink. In wood that's continually exposed to the weather, shrinkage can be considerable. Try to buy kiln-dried lumber, even if this means purchasing from a supplier other than the one you usually use. If dry wood is not available, or if its added cost is not in your budget, at least make sure that the moisture content is consistent throughout your selection. You may not be able to prevent shrinkage, but if you plan for it ahead of time the deck will look better because the gaps between the boards will look uniform. Different-size gaps will make the work look sloppy.

The grade and species of lumber you select will directly affect the longevity of your deck. In parts of the West, decks are frequently built from cedar or redwood. These species are readily available and quite resistant to decay. But in the eastern part of the country, pressure-treated lumber is used most often because it withstands our harsh climate, and is generally more available and less expensive than cedar or redwood.

There are two grades of pressure-treated lumber suitable for decks. We strongly recommend using #1 yellow pine, particularly for the rail-

Decks ease the transition between the house and the landscape, and also serve as an entry. Decks should be functional, durable, well proportioned and attractive. Properly designed so they won't trap water, they will withstand the destructive forces of the weather.

ings, benches and decking. The quality of #1 pressure-treated lumber is fairly consistent, and the material is easier to work than #2 grade. The #2 grade has a greater number of open knots, and these weaken the boards and encourage water to accumulate. Large knots that span more than half the width of a board are very dangerous in either grade, since the pressure of a footstep or rough handling during construction will sometimes snap the board in half. One problem with pressure-treated lumber is warpage. It can twist severely, cup and bow if not handled correctly. Keep it covered and out of the sun until you use it.

Pressure-treated lumber often has a greenish color, due to the chemicals it's impregnated with (usually chromated copper arsenate). This tint will weather away into a pleasing light grey in about two years, though the treatment chemicals still protect the wood. Some people want more color to their deck, however, so we recommend a semi-transparent stain. If you wait a year or so before the first application, the wood will have a chance to dry out and will accept a fuller coating, doubling the stain's expected life. We prefer stain to paint because paint traps moisture and requires more maintenance.

When you order the lumber for a deck, include about 10% more than you think you'll need. This will prevent time-consuming trips to the lumberyard if your estimate was slightly off, and allow you to cull out badly warped boards with too many knots.

When the lumber is delivered, remember that moist lawns and delivery trucks are a bad combination. There are better ways to find out where the septic-system drainfield is than to have a truck crush the drain tiles. And since the chemicals in pressure-treated lumber can kill grass, make sure you relocate lumber piles after three days to a different location on the lawn.

Nails—We use only galvanized nails on deck projects, 10d for the decking and 16d for framing. Two types of galvanized nails are available. Electro-plated nails have a smooth finish and take less effort to pound in, but hot-dipped nails, with their rough surface, grip much better and are also more rust-resistant. To save time in laying down decking, we use a pneumatic nailer and resin-coated galvanized nails. The resin coating heats up when the nail penetrates the wood, and then hardens like glue for a firm grip.

If you have problems with lumber splitting as you nail into it, use your hammer to blunt the end of your nails. This way they will puncture the wood instead of piercing and splitting it.

Piers—Like the foundation of a house, the foundation of a deck must transfer loads from the structure to the ground. But unlike the foundations of most houses, deck foundations are not continuous. To support the deck, a system of concrete piers is used. The piers extend from grade level to below the frost line—32 in. to 48 in. in our climate. A pier that does not go below the frost line will eventually heave and push the deck out of level.

The standard method of determining where the piers will go requires string, a collection of

stakes, and the application of some basic practical geometry. This method works well on decks with simple rectilinear froms, but complex forms are considerably trickier to deal with. When we are faced with the task of building elaborate forms, we've found a way to locate piers that works quite well, and that allows for design flexibility as the project progresses.

Rather than spend a lot of time and effort to locate all the piers at once, we use a locate-build-locate process. The idea is to define the limits of the deck, locate and then pour perimeter piers. Once this is done we can frame the perimeter of the deck, bracing it in place. After that, we locate the rest of the piers.

The easiest piers to locate are the ones that must be placed at a particular point. If you know, for example, that you want the edge of the deck to change direction about 10 ft. from the house and 14 ft. from the oak tree, dig and pour a pier there. The process is empirical: you build what you know in order to answer questions about what you don't know.

After the concrete has partially cured in about 24 hours (it will take nearly a month to gain most of its strength), we begin framing. This method may seem somewhat backward, but we often find it much easier and more accurate in the long run to dig some piers to support interior spans after the perimeter is established.

You will need a long-handled shovel, a digging bar and a post-hole digger to dig the holes for piers. There are two kinds of manually operated post-hole diggers, and you may end up using both of them on your deck project. A post-hole auger looks and works somewhat like a giant corkscrew; as you turn it into the earth it pulls dirt from the hole. An auger works particularly well in hard ground, but is easily stymied by rocks. A clamshell post-hole digger looks like two long-handled spades hinged together at the ferrule, with the blades opposing each other. The work goes quickly in soft ground, but more slowly in packed or clay soils. The clamshell is less likely to stall when you hit rocks, since it can reach into a hole to remove them, but large rocks can cause problems.

If you have a lot of holes to dig, you can rent a gasoline-powered hole digger. This is basically a power auger, and we prefer the one-person model with a torque bar because it won't take you for a ride when it hits a rock.

A long digging bar comes in handy for loosening dirt and breaking rocks that can't easily be removed from the hole in one piece. It also helps loosen tightly packed soil. This solid, heavy, steel persuader is pointed on one end, and can also be used to pry out rocks.

Rocks are the main problem in digging footings around here, but roots can also be a nuisance since decks are frequently near large trees. Use an old handsaw or sharp ax to cut the roots cleanly, but don't seal the cut ends. A botanist once told me that a root or branch will heal itself, and that tar and other sealants interfere with this process.

The shape of the holes you dig is nearly as important as their depth. They should generally be round, and about 8 in. to 12 in. in diameter. The sides of the hole should be reasonably

smooth, and the bottom of the hole should be slightly larger than the top to distribute loads well. If there are any ledges or if the hole narrows at the bottom, the freeze/thaw cycle will lift or tip the pier as much as 12 in. over time. Make sure that the floor of the hole is undisturbed earth, because a layer of soft earth here will allow the pier to sink.

We usually pour concrete directly into the hole, using the sides of the hole to form the pier. You can also use Sonotubes to line the hole. These cardboard tubes, available in various diameters from masonry-supply stores, are especially handy if you want the concrete to extend above grade to form a pier. If you suspend the tube 6 in. above the bottom of the hole when you pour, the concrete will ooze out the bottom to widen the base of the pier and increase its bearing ability. Piers should include #4 reinforcing bar if they extend more than 6 in. above grade.

A good concrete mix for piers is 1:2:3, which means one part portland cement, two parts sand, and three parts gravel (¾-in. or 1-in. gravel will be fine). An alternative to mixing your own concrete is to purchase ready-mix, which is a pre-proportioned cement, gravel and sand mixture that usually comes in 90-lb. bags. The portland cement will sometimes settle to the bottom of ready-mix bags, so it's a good idea to dry-mix the contents of each bag before adding water. A wheelbarrow is great for mixing concrete in, but be sure to wash it out afterwards, along with your mixing tools.

Finding level—Building a deck can be an exercise in elementary civil engineering, and many beginners are frustrated by having to find the proper relationship between posts and boards that aren't connected. You can't always use a carpenter's level to do this—how would you check two posts, 15 ft. apart, to see if they are at the same height? We often use a 2-ft. level on a long, straight board to check for level, but other tools can be used as well.

A string level is a small spirit level that hooks onto a length of layout twine. When the twine is pulled taut, a rough estimate of level can be determined by raising or lowering one end of the twine and watching the bubble in the level.

An optical pocket level is something of a cross between a telescope and a transit. Looking through it, you align a small leveling bubble with cross hairs to determine an approximately level visual line.

A water level is an inexpensive and very accurate homemade device used to check the relative heights of widely separated items. It's made from clear plastic tubing filled with water (a few drops of food coloring will make the water easier to see). Because of atmospheric pressure, the water level at one end of the tubing always matches the water level at the opposite end, no matter how many twists and turns the tubing takes. It's particularly useful over long distances, as when you want to compare the heights of ledger and posts.

A transit is a precision instrument used by surveyors, and this is what we use to determine level, plumb and the relative heights of widely distant objects with a high degree of accuracy. The transit is fairly expensive, but if you do a lot of decks, the money is well spent.

The ledger—The ledger is a length of 2x lumber that is attached directly to the house, allowing a portion of the deck to "borrow" the foundation of the house for support. It's usually the first framing member to be installed and should be selected from the straightest stock available, since it serves as a reference point for much of the work to follow. When you install the ledger, don't rely on siding or the foundation to be level, because often they're not.

The top of the ledger supports the decking, and if you think of it as a rim joist, you'll get the idea. If the ledger has to be attached to the house foundation in order for the final deck elevation to be where you want it, you'll have to fasten it with lag bolts and lead expansion shields or some other masonry-anchor system. Masonry nails won't work very well, particularly in poured foundations that have had many years to cure. When using a standard masonry bit in an electric drill to bore holes for the expansion shields in a concrete foundation, use a star drill to break apart any pieces of aggregate you can't drill through. We've found that a roto-hammer speeds this job considerably.

Sometimes the plans will call for the ledger to be fastened above the house foundation, and in this case, 4-in. by ⅜-in. galvanized lag bolts spaced about 24 in. apart and fastened to studs or a rim joist will usually do the job. Slide flashing under the existing siding and over the ledger, if possible, to keep water from seeping behind it. If the ledger is going to be mounted to some sort of concrete patio or walkway, use shims to hold the board away from the concrete to allow the water to pass freely by.

Allow at least 1 in. between door sills and the decking surface to prevent any water from running back off the deck into the house. If you include a step, use the same step rise used elsewhere on the deck for the sake of consistency.

Posts—Posts transfer the loads from the deck structure to the piers. One end of each post is attached to a beam or a joist, and the other end rests on the pier. We usually don't anchor posts to the piers, since the weight of the deck is enough to keep them in position. Though many people embed posts in concrete, we feel this technique can create serious problems if water collects between the post and the concrete. Posts are usually 4x4s, 4x6s or 6x6s. A 6x6 post is generally more than needed for bearing purposes, but it enables us to notch beams or joists into it for added strength.

Beams—Beams are an intermediate structural member, used to support joists. They can be solid lumber, usually 4x6 or 4x8, or they can be built up out of 2x lumber. When you're fabricating built-up beams, a common mistake is to nail the individual 2x material face to face, which allows water to get trapped between these boards. Instead, you should sandwich blocks of ½-in. pressure-treated wood between the boards to create a void for water to run through.

Joists—Joists are the uppermost structural element supporting the decking. They are generally 2x lumber, and should be reasonably straight. When laying joists into place, make sure that any crown in the board is facing up; in time, gravity and the weight of the decking will straighten the joists.

There are at least two good techniques for attaching joists to the ledger, and at least one that should not be used. One reason many decks decay at this location is that the joists are toenailed into the ledger; nails split the ends of the joists, allowing water to collect exactly where it shouldn't. Joist hangers minimize splitting at the joist end, and will prevent water from getting trapped in this crucial joint. We first nail the hangers to the joists, and then fit the assembly to a level chalkline. This compensates for slight variations in joist width.

A cleat can also be used to support the ends of the joists. You still need to toenail the joists, but splitting is reduced because the nails can be a smaller size since they do not carry the weight of the deck. If you use this technique, the ledger should be one dimension wider than the joists. For example, use a 2x8 ledger with 2x6 joists. This will allow room for a 2x2 cleat to be attached to the ledger. Run a bead of silicone caulk along the cleat/ledger seam to keep the water out.

Decking—If the decking surface is not applied properly, it will be the first thing to deteriorate, causing a chain reaction of decay throughout

Anatomy of a deck

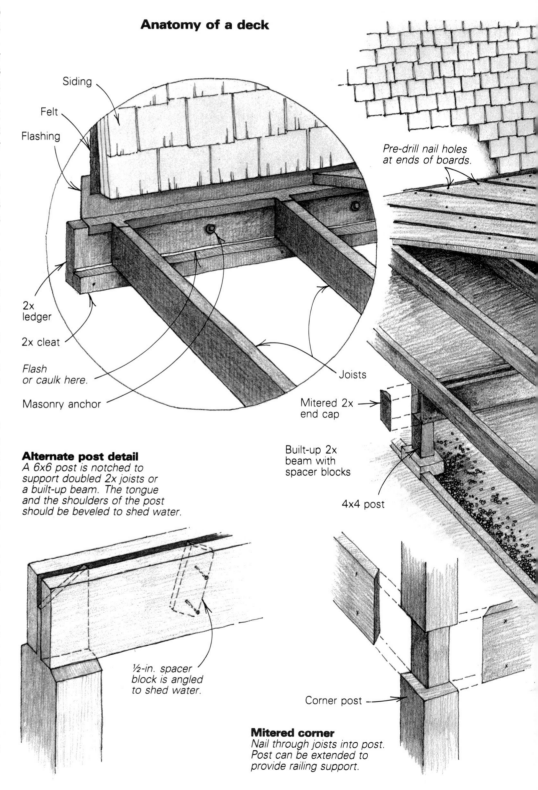

Siding

Felt

Flashing

Pre-drill nail holes at ends of boards.

2x ledger

2x cleat

Flash or caulk here.

Masonry anchor

Joists

Mitered 2x end cap

Built-up 2x beam with spacer blocks

4x4 post

Alternate post detail
A 6x6 post is notched to support doubled 2x joists or a built-up beam. The tongue and the shoulders of the post should be beveled to shed water.

½-in. spacer block is angled to shed water.

Corner post

Mitered corner
Nail through joists into post. Post can be extended to provide railing support.

Drawing: Frances Ashforth

the rest of the structure. *Never* butt the ends of two decking members together and nail into a single joist. This is one of the major causes of decking failure (top photo, next page). The reason is that water collects in the seam between the butting boards and enters their end grain. And when two boards butt over a single joist, the problem is intensified. With only ¾ in. for each deck board to be nailed into, the nails must be placed very close to the end of the board, encouraging splitting. We get around these problems by doubling up strategic joists, using a block of wood between them to create a 1½-in. space. The end of each deck board cantilevers over this space so water can't collect. This also allows the boards to be nailed farther from their ends, which minimizes splitting. If

Bench supports built from 2x8s can be trimmed to a width of about 3½ in. where they form the backrest. The supports should be securely nailed or bolted to the deck's supporting structure.

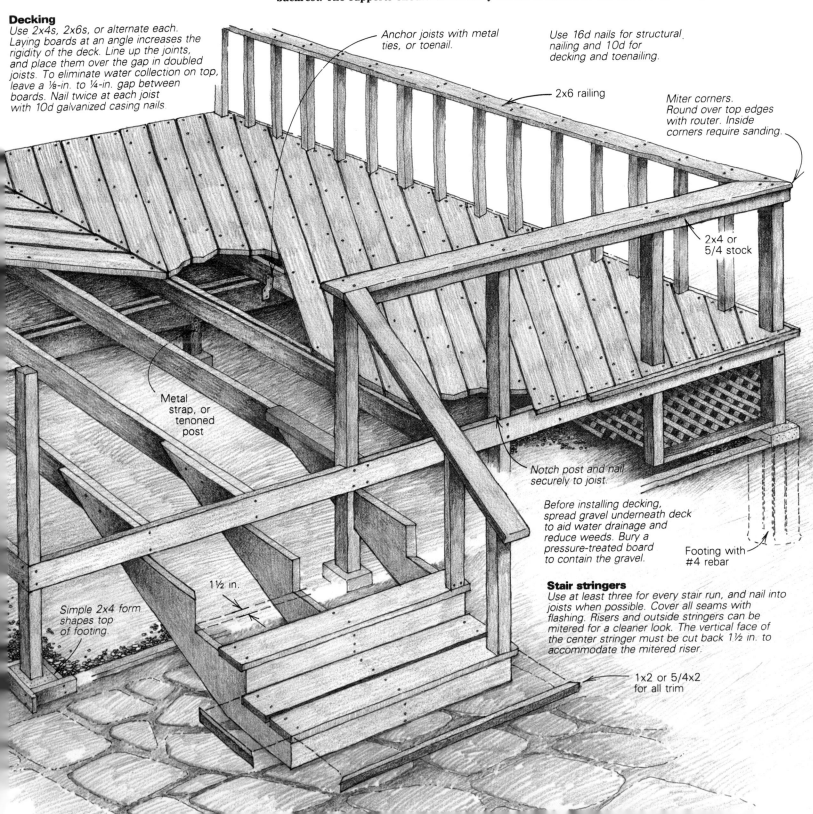

Decking
Use 2x4s, 2x6s, or alternate each. Laying boards at an angle increases the rigidity of the deck. Line up the joints, and place them over the gap in doubled joists. To eliminate water collection on top, leave a ⅛-in. to ¼-in. gap between boards. Nail twice at each joist with 10d galvanized casing nails.

Anchor joists with metal ties, or toenail.

Use 16d nails for structural nailing and 10d for decking and toenailing.

2x6 railing

Miter corners. Round over top edges with router. Inside corners require sanding.

2x4 or 5/4 stock

Metal strap, or tenoned post

Notch post and nail securely to joist.

Before installing decking, spread gravel underneath deck to aid water drainage and reduce weeds. Bury a pressure-treated board to contain the gravel.

Footing with #4 rebar

Simple 2x4 form shapes top of footing.

1½ in.

Stair stringers
Use at least three for every stair run, and nail into joists when possible. Cover all seams with flashing. Risers and outside stringers can be mitered for a cleaner look. The vertical face of the center stringer must be cut back 1½ in. to accommodate the mitered riser.

1x2 or 5/4x2 for all trim

The premature failure of decking is often caused by nailing too close to the end of the board. This encourages splitting, and allows water to accumulate around the board ends.

Cutting the decking to length after nailing it in place ensures a clean, uniform edge and saves time otherwise spent cutting boards one by one. Run the boards long and snap a chalkline to mark the cutting path, or use a straightedge to guide the shoe of the saw. A built-up beam and a mitered joist corner can be seen in this photo. Wherever joists along the perimeter of the deck meet, a mitered connection protects end grain from direct exposure to the weather.

you must nail closer than 2 in. to the end of a board, predrilling the nail holes can also help to reduce splitting.

For the decking surface, we like to use 2x6s or sometimes 5/4 by 6s if the quality is good. You can also use 2x4s, but you'll have a lot more nailing to do. If you have the choice, nail the decking cup-side down to prevent any water from pooling on the individual boards.

We like to space the decking boards about 1/8 in. apart (the thickness of a 10d nail). The boards will shrink, depending on their moisture content, and we have seen this space expand up to 1/4 in. If there are a lot of deciduous trees with small leaves close to the deck, you might want a wider spacing to allow the leaves to fall through. When you're installing the decking, some boards will most likely be crooked. With a flat bar and some lever action, one person can easily straighten out each board while nailing.

Stairways—Stairways can be dangerous areas, and require special attention. Codes usually call for at least one railing at the side of the stairs if they include more than two risers. Wide stairways have a spacious and inviting appearance, so we like to build them at least 4 ft. wide, enough for two people to pass comfortably.

Inconsistency in the height of a step or length of a tread is dangerous and awkward. We have found that a shorter rise and longer tread is easier to walk up, safer and very elegant. A rise of about 7 in. seems to be make a comfortable step, and allows us to use an untrimmed 2x6 for the risers. Sometimes we miter the riser boards into the outside stringers. We're not fond of exposed stringers. We use a pair of 2x6s for the tread; these make a run of a bit more than 12 in. with a trim board.

We use three stringers for deck stairs, even if the stairway is only 30 in. wide. We have seen too many stairways warp and fall apart with only two stringers. Using three-stringer construction isn't too difficult and doesn't cost much more. In fact, the trickiest part of building one is cutting each stringer to identical shape and lining them all up in the same plane on the rise and run. You can buy framing clips (called stair-gauge buttons) to keep your framing square in the right position while you lay out each stringer, but this doesn't always get you past Murphy's law. Our trick is to lay out one stringer only and clamp it to another length of stringer stock. When you cut the first one, the sawblade scores the second one, saving you one layout. Repeat the process with additional stringers. This will duplicate all the stringers and save layout time.

Railings—Railings are important to the safety and appearance of decks. As a design element, they can be the highlight and outline of your project. Railings are one of the places where a designer's creativity can be expressed, and there is no one way to build them. But there are some general rules to follow.

The most important characteristic of a good railing is strength. You can be sure that people will lean against the railing, and often they will sit on it, too. A strong railing is particularly important on elevated decks. We like to play it

safe, and build our railings as strong as possible. To do this, we solidly attach the posts or balusters to the deck joists. These structural supports should be located no more than 48 in. apart, and the railing should be about 34 in. above the decking surface. Be sure that the edges of the railing are well sanded, especially in places that will get a lot of use, like stair railings.

Consider the spacing between your railing uprights as part of the project's visual design. Close spacing visually encloses the space, and also prevents small children from falling through. Railings with fewer uprights will visually expand the space and be less inhibiting to your view.

One type of railing we use combines a bevel-cut 2x6 and a matching 2x4 into an "L" shape. We position the 2x6 horizontally, and our clients enjoy the strong visual effect this creates. The 2x6 acts as a cap for the railing uprights, shielding their end grain from the elements. Although tricky, this technique allows for some very interesting joinery at all corners. To clean up the mitered edge, we round it over with a router or belt sander. A simpler railing is shown in the drawing, previous page.

Seating—Built-in seating is a great way to finish the deck. As with railings, there are many ways to build it, and no one way is correct.

You can build seating either with or without a backrest, and the choice will often depend on whether or not an unobstructed view is important. If comfort is more important, you'll want to build at least some of the seats with a backrest. The top of the backrest should be between 30 in. and 34 in. from the deck, which can be designed nicely to tie in with the railing.

A backless bench can function as a physical barrier for a deck edge without acting as a visual barrier as well. We have also used a low, wide railing as a mini-bench, which also makes a good place to display potted plants. And sometimes we'll use a built-in planter to serve as a visual barrier.

Benches can be difficult to build, since they not only have to be strong, but comfortable too. With backless benches, we have used 4x4s or 2x6s as supports. For a bench with a backrest, we use a 2x8 as the seat support, and rip it down to a 2x4 for the backrest support (photo previous page). A 15° backward lean seems to be comfortable. We then run our mitered railing across the top at 32 in. The cross supports for the seat are 2x6s cut into long, wide triangles. For the seat, we use three 2x6s with a 2x4 band. This will give the seat a total width of 18 in. The standard seat height that we use is 17 in. In calculating seating height, don't forget to account for any cushions that you might use. Save your best boards for construction of the seating, because this part of the deck will be well used and very visible.

A final note—Use these tips in combination with your local codes. With a little creativity and your basic construction knowledge, the deck you build should last a long time. □

Scott Grove is a partner in Effective Design, a design/build company in Rochester, N. Y.

Photos: Scott Grove

A Screened-porch Addition

Simple detailing and inexpensive materials provide a shelter from the swarm

by Jerry Germer

Screened porches have graced our homes since horse tails were first woven into screens in the mid 1800s. On hot summer evenings, people would sit on the porch, perhaps in a swing hung from the roof, and pass the time with family, friends and neighbors. But in the middle of this century,

Roof structure. A three-tiered 15-in. fascia conceals a very shallow-pitched shed roof constructed of pressure-treated framing and corrugated, fiber-reinforced plastic (FRP) panels.

ing to negotiate steps or pass through the living room. Steps would lead from just outside the porch deck to the lawn.

To economize on materials, I decided to lay out the plan on a 3-ft. by 3-ft. grid, yielding a 15-ft. by 9-ft. deck floor—just about right for barbecue equipment, a small table and a few chairs. An existing window and the wall section below it would be cut out of the living-room wall to provide a passageway to the porch from indoors. A screen door would link the porch to the stairs and to the deck/walkway.

Getting the deck to float—The bank at the rear of the house drops off about 5 ft. right away. If I built the new porch at the same level as the living-room floor, the open space below the porch would range from 1 ft. to 4 ft. Closing off this space with open, lattice-type skirting would have been tricky, and a solid foundation seemed an even less attractive alternative—I didn't want to stand in the backyard and see a 4-ft. high wall of parged concrete block.

A simpler and more elegant solution was to cantilever the floor structure over the sloping bank, supporting the joists on a perpendicular 4x10 beam 6 ft. from the house (drawing p. 59). The beam rests atop two 4x4 posts, which bear on concrete footings resting about 3 ft. below grade. The posts are spaced 9 ft. apart, so the beam cantilevers out 3 ft. at each end. I used 2x8s on 3-ft. centers for floor joists, fastening one end of each joist (with a joist hanger) to a 2x8 ledger bolted to the house sill. The floor of the porch consists of 2x4 boards in continuous lengths, spaced ½ in. apart and screwed to the joists. All structural members, as well as the decking, are pressure-treated Southern yellow pine. The effect, with the house's skirtboard wrapping around the joists, is of a porch that floats out over the bank.

With no foundation walls to which I could attach screen, though, I needed to find another way of maintaining continuity of the insect barrier. The only solution that came to mind was to screen the floor. That's why I stapled fiberglass screen over the tops of the joists, before screwing down the decking. It worked; bugs can't get up through the bottom of the deck. But after two years of summertime use, the porch revealed the flaw in my scheme. Bits of debris, ranging from dust to pencils, fall into the cracks and get trapped by the screen. Vacuuming has been only partially effective in cleaning out the cracks, but then again, our aging Electrolux no longer has the suck it once did. Nevertheless, if ever I have occasion to build another screened porch, I'll take the to come up with a solution that allows for periodic cleaning of the screen.

Daylight and privacy—I wanted the walls of the porch to be as light and open as possible, yet still offer some privacy. Because the roof structure would be very lightweight, supporting posts were kept lean as well—2x4s on 3-ft. centers. Two posts meet at the corners so that neither post overlaps the inside face of the other. I stapled 36-in. widths of black fiber-

when postwar builders were faced with the need to build massive numbers of houses quickly and economically, they built smaller houses on smaller lots. Large outside porches no longer fit the houses or the lots. Lifestyles changed, too: People were spending more of their free time inside and in front of the television.

Recently, however, screened porches have been making a quiet comeback. Has television lost its charm, or are people rediscovering the outdoors? I'm not sure, but for our family, a screened porch (photo previous page) was the only way we could enjoy the

beauty of our rural New Hampshire site without being assaulted by hordes of pesky insects.

Making plans—The north side of the house off the living room seemed the obvious location for the new porch. Snugged against the house on its shadiest and most private side, the porch could project over the small bank that drops away to the backyard. A 3-ft. wide walkway could run along the back wall of the house, connecting the porch with the back door near the kitchen. That way, we'd be able to bring food out onto the porch without hav-

glass insect screen to the inside face of each post, then secured 1x2 wood strips over the stapled edges with drywall screws.

Black screen over regularly spaced posts doesn't make for a very interesting wall surface, however, nor does it provide much in the way of privacy. Hoping to take care of both of these problems, I wrapped 30-in. high panels of 1x2 balusters around three sides of the porch. The panels are screwed to the outside faces of the posts, which allows removal of the panels for maintenance and repainting. The effect though, with all balusters spaced approximately on 5-in. centers, is of a continuous rail.

Keeping rain out, letting light in—Most roofed porches have a downside. In providing a sheltered space outdoors, they shade the windows, darkening the rooms. While summers here are short, winters are long, and gray days abound. When I remodeled our house, I tried to maximize passive-solar gain. By adding more windows on the south side of the house and removing non-bearing interior partitions on the first floor, I'd been able both to warm the house and to make it feel lighter and more cheerful. But in my zeal for energy efficiency, I had also eliminated some north-facing windows. I didn't want to block out light remaining from the north windows by shading them with an opaque porch roof.

"Why not design a roof that would allow light to pass through?" I thought to myself. If I could find the right material, the idea might have promise. Glass was ruled out immediately as being too expensive. Options in plastic included double-skinned polycarbonate sheet (such as Exolite) and corrugated fiber-reinforced plastic (FRP). Exolite would work but isn't cheap. FRP would be cheap, but in order to drain properly, the panels would have to overhang the eave. The exposed ends, undulating like a washboard, would fit better on a shed or a chicken coop—not at all in keeping with the character of our house. If I were to use the FRP, I would have to come up with an eave detail that hid the corrugated panel ends without impeding drainage.

A shallow pitch concealed—The two challenges confronting me were how to make the porch roof seem to belong to the house and how to hide the corrugated ends of the FRP from view. A shed roof that matched the pitch of the house's roof would run smack into the second-floor windows. And a shallower pitched roof would seem an afterthought. A completely level roof, on the other hand, would underscore the house's horizontal eave and frieze board. But I still needed to provide some slope for

Behind the fascia. Vertical cleats nailed to the joist support the fascia. Flashing and an angled 1x6 protect the fascia boards from runoff.

Section through porch

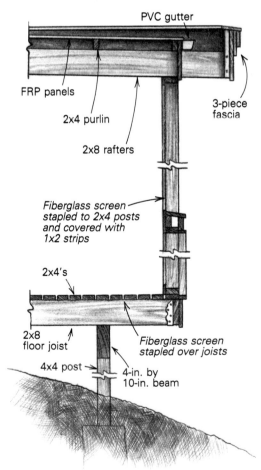

PVC gutter

FRP panels

2x4 purlin

2x8 rafters

3-piece fascia

Fiberglass screen stapled to 2x4 posts and covered with 1x2 strips

2x4's

2x8 floor joist

4x4 post

Fiberglass screen stapled over joists

4-in. by 10-in. beam

drainage. A solution was suggested by the roofing material itself.

Because the panels are 12 ft. long, I'd be able to use them full-length (no horizontal lap joints), all but eliminating the chance of wind-driven rain getting up under my roofing. I figured I could get by with a minimum pitch for drainage—say, ¼ in. per ft. Later, I could wrap a fascia around the three exposed sides of the roof to hide the rafters, the result being a flat-looking roof with good drainage (photos left and facing page).

Because my goal was to let in as much light as possible, I left all roof framing exposed. The 2x8 rafters, spaced 3 ft. o. c., run perpendicular to the house, dropping 3 in. in 12 ft. The high ends of the rafters are fastened to a 2x8 ledger lag-bolted to the house. The lower ends of the rafters extend 15 in. past the top of the outer wall, to carry the fascia. I also ran 2x4 purlins (22 in. o. c.) 15 in. out to support the fascia at the side walls. Two ⅛-in. cables run diagonally under the rafters, corner to corner, to brace the roof against racking. The FRP roofing panels were then attached to the purlins with aluminum roofing nails and rubber washers. The nails are spaced 6 in. to 8 in. o. c. (I had to predrill the panels).

Details, details—Concealing the rafters and the FRP would require a 15-in. wide fascia, which, I felt, would be overly heavy-looking if installed in one piece. Some horizontal lines would be necessary to reduce the apparent width of the fascia and to add a little interest to an otherwise plain façade. So I built a three-leveled fascia, with layered boards of diminishing size.

Behind the fascia, the FRP panels project over the porch's front wall by about an inch. A PVC gutter attached to an interior fascia above the screen wall collects runoff and carries it to downspout tees at each end (photo above). Rather than run downspouts down the corners of the porch, where they would have messed up the corner-post detail, I elected to let the water drip directly to the ground.

The translucent roof panels function well in winter, allowing a great deal of northern light into the house. In the summer, the porch is shaded by the house throughout the day, except for early in the morning and late in the afternoon. But the late afternoon summer sun was a nuisance. Hoping to resolve the problem simply and inexpensively, I draped sheets of burlap on 1x1 battens from eyehooks screwed into the purlins, a solution that has worked quite nicely. According to my wife, an architect never knows when to quit. □

Jerry Germer is an architect and writer living in Marlborough, N. H. Photos by Vincent Laurence.

Controlling Moisture in Deck Lumber

Many problems associated with deck deterioration can be traced to the original moisture content of the lumber

by Bob Falk, Kent McDonald and Jerry Winandy

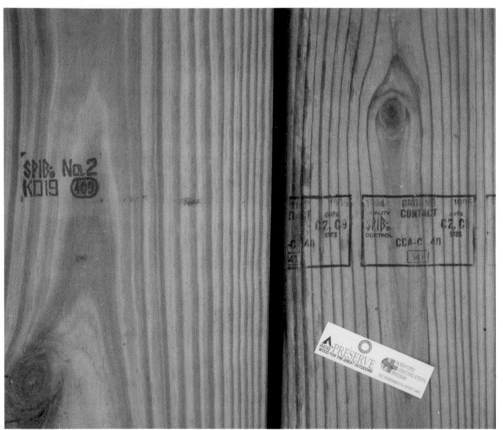

Look for the mark to be sure what you're getting. Grade marks, such as the ones on these two pieces of pressure-treated lumber, indicate the maximum moisture content of the wood, which is information that's good to know if you're going to use the wood to build a deck. Some pressure-treated wood will carry two marks, one from the mill and one from the pressure treater.

Properly dried lumber is worth the trouble. To dry lumber properly, align the stickers, or spacer strips of wood, over pile supports to promote even weight distribution and optimal drying. Place weights on top of the stack of wood to stop the top boards from warping.

If you've ever inspected, repaired or torn off an old wood deck, you know what can go wrong with one. Nail heads stick up. Deck boards decay, cup or twist, and joints that once were tight open up and loosen. Bad construction, the use of unsuitable lumber, the wrong fasteners or a lack of maintenance are often the sources of the problems. However, there is another important factor that can affect deck performance.

Here at the United States Department of Agriculture Forest Service Forest Products Laboratory in Madison, Wisconsin, we've learned from extensive research into wood behavior that the origin of many of these problems often can be traced to the moisture content (MC) of the wood at the time the deck was built or to the effects of moisture during its lifetime.

The effects of moisture in deck lumber determine how good a deck will look, how well it will hold up and, often, how long it will last. Obviously, it's impossible to control the amount of humidity and rain a deck is exposed to. (You can limit the amount of moisture that comes in contact with the wood only by applying a proper water-repellent finish or by purchasing lumber that has a water-repellent finish.) However, you can control the amount of moisture in the wood. Too much—or even too little—moisture in wood eventually can lead to structural problems.

Moisture content can affect a deck for years to come—To minimize warping, splitting, checking, shrinking and failing finish, the deck boards at the time of construction should be uniform and less than about 20% MC, regardless of the species or whether the wood has been pressure-treated (top photo).

In most areas of the United States, we expect lumber in aboveground, protected, exterior applications to reach an equilibrium moisture content (EMC) around 12%. If your specific site is normally either very wet or very dry, the EMC will be higher or lower, respectively.

In general, the moisture content of most treated lumber is high—in the 35% to 75% MC range—and the wood is still wet when it arrives at the job site, unless it has been kiln-dried after treatment and marked KDAT. If the wood is stamped KDAT, its moisture content should be about 19% or less. Because redwood and cedar aren't treated with preservatives, they're usually marketed as kiln-dried or as air-seasoned, which means they will have about a 20% MC. Most deck builders install deck boards on delivery. Although this way is easiest, pressure-treated boards probably will vary greatly in moisture content and often will shrink unevenly.

In the case of preservative-treated wood, we recommend KDAT lumber, when available, because many problems that eventually surface in deck construction are a result of using wet lumber. Another option is to air-dry the treated lumber yourself. In both cases, you'll be able to identify problem deck boards before installation and exclude them from your project.

Air-dry pressure-treated lumber to equalize moisture content—Treated lumber that's not marked KDAT should be air-dried for several weeks, depending on the type of weather and the extent to which the lumber is exposed.

Usually, pressure-treated wood comes directly from the treater and is bound and shipped wet to the lumberyard, where it often is stored outside and unprotected. Air-drying for several weeks will help even out the moisture-content differences between the pieces of wood and, on average, will lead to a more consistent moisture content at installation.

In the long run, it is worthwhile to order the lumber to arrive at the job site a few weeks early to allow time for air-drying. Air-drying also is recommended if you build a deck with redwood or cedar that contains a moisture content much greater than 20%.

The air-drying method we recommend is stacking the lumber in layers separated by narrow strips of wood, or stickers, to allow air to move freely between layers (bottom photo, facing page). Care should be taken to align the stickers vertically within the pile. Alignment helps to distribute the load evenly and to minimize warping during drying. Also, it's a good idea to place weights, such as concrete blocks, on top of the pile to help minimize twisting of lumber during drying. (Avoid iron weights because they can stain the wood if they get wet.)

If the pile is protected from the weather—either by a shed or by plastic sheeting—and is allowed to dry several weeks, the lumber should reach a moisture content of close to 20%.

Using an electronic moisture meter is a simple method of measuring moisture content in wood. This kind of meter typically measures the electrical resistance between two metal pins driven into the wood. Moisture meters must be calibrated depending on wood species and temperature. You can purchase a moisture meter for around $100.

Moisture content also can be determined by weighing a few representative small pieces of wood, drying the pieces in an oven at 200°F for 24 hours and weighing them again after they're oven-dry. Divide the difference between the original weight and the oven-dry weight by the oven-dry weight, then multiply by 100 to get the moisture content in the form of a percentage.

Shrinking can be used to your benefit— Wood shrinks only when moisture content falls below about 30%. A 6-in. wide treated southern

pine deck board should shrink by about ³⁄₁₆ in. if it reaches 12% EMC, so laying wet decking boards tightly against each other should result in a ³⁄₁₆-in. gap when the boards dry (photo top right). For redwood or cedar purchased at 20% MC, a nominal 6-in. decking board will shrink only about ¹⁄₁₀ in. when a 12% EMC is reached. If the lumber installed is drier than the local EMC, and if the boards are laid tight, there's potential for the wood to pick up moisture, swell and buckle.

Depending on the lumber species and moisture content—as well as the desired gap between boards—a gap between deck boards can be planned based on the amount of expected shrinkage. We suggest a final gap of about ³⁄₁₆ in. to ¼ in.—not big enough to catch a small heel, but big enough to allow dirt, leaves and other debris to fall through.

Warping and cupping (photo center right) usually are caused by uneven shrinkage between the top surface and the bottom surface of deck lumber. The cupping of individual boards is aggravated because the top surface is usually at a lower moisture content—because of exposure to the sun and wind—compared with the protected bottom surface. This situation means that deck boards installed wet are likely to warp the most, especially when installed during hot months. This shrinkage difference is more pronounced if the bottoms of the boards remain damp, such as when the deck is built low to the ground or near wet soil.

Finally, we don't recommend using deck boards wider than 6 in. because cupping and warping can become excessive.

Don't forget to finish the job—Even though you're using naturally decay-resistant or pressure-treated wood, the horizontal surface of a deck is exposed to foot traffic, sun and rain, which makes finishing a deck with a water-repellent preservative a necessity. This exposure will degrade the wood's surface, and unless the wood receives the proper finish, discoloration and checking often result, leading to a rough, uneven deck surface and decay in untreated wood (photo bottom right). Applying and maintaining a finish on your deck will help minimize problems.

It used to be recommended to wait a year, or a season, to finish a deck. In our experience, this amount of time is too long because surface problems that cannot be corrected later may develop (i. e., checking, cracking, splintering).

For a new deck, apply the finish after the surface of the wood has dried to about 20% MC. Wood stamped S-DRY, KD (kiln-dried), MC-15 (average moisture content 15%) or KDAT has been dried and can be finished immediately. □

Bob Falk, P.E., is a structural engineer and Kent McDonald and Jerry Winandy are wood scientists at the USDA Forest Products Laboratory in Madison, Wis. They are coauthors of "Wood Decks: Materials, Construction, and Finishing," a handbook published in cooperation with the Forest Products Society (2801 Marshall Court, Madison, Wis. 53705-2295; 608-231-1361). Photos by Steve Schmieding, Jim Vargo and the authors except where noted.

Gaps between boards shouldn't be this wide. The gaps between these deck boards are much too great and can result in injury for women wearing high-heeled shoes or for small children. Gaps this wide can result when wet boards are laid with a gap between them. The gap becomes larger as the boards shrink.

Evenly dried, properly fastened boards resist warping. Warping and cupping are usually caused by uneven shrinkage between the top surface and the bottom surface of wood. The wrong types or placement of fasteners also can cause wood to warp and cup, as shown above. Photo courtesy of the Southern Forest Products Association.

All decks need regular maintenance. Even cedar, redwood and pressure-treated lumber require regular application of a water-repellent preservative (including mildewcide). But lumber that is not naturally durable or that is not preservative-treated will quickly decay, resulting in a short-lived and vulnerable deck. Note the water intrusion around the long surface check of the leftmost deck board. In a year or two, this board will degrade to the same condition as the center board.

Details for a Lasting Deck

Government scientists study outdoor structures and report on which details, fasteners and finishes hold up best

by Bob Falk and Sam Williams

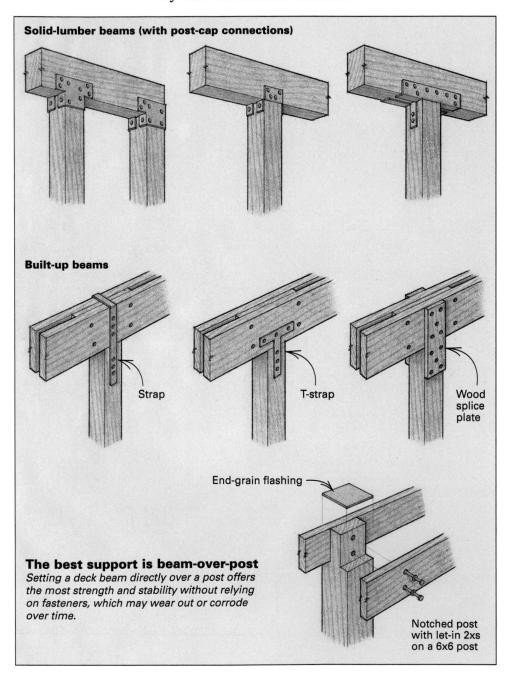

Solid-lumber beams (with post-cap connections)

Built-up beams

Strap

T-strap

Wood splice plate

End-grain flashing

The best support is beam-over-post
Setting a deck beam directly over a post offers the most strength and stability without relying on fasteners, which may wear out or corrode over time.

Notched post with let-in 2xs on a 6x6 post

Some decks need major overhauls after less than 10 years. Others stay strong and good looking for decades. What's the secret? Well, besides the obvious first choice of suitable lumber (we recommend either a naturally durable species or preservative-treated lumber), a lasting deck is put together with strong, durable fasteners, and it gets regular applications of a penetrating finish to repel moisture and to minimize the effects of the weather.

Although the structure of a deck is a lot like the skeleton of a conventionally framed wood house, a deck doesn't have the stability of sheathing, and there's no roofing and siding to protect it from the elements. That's why decks require extra care and attention to detail. As wood researchers at the U. S. Forest Service Forest Products Laboratory, my colleagues and I have studied lumber, construction techniques, fasteners and finishes. From this research, we offer some recommendations for building decks that last.

Start with good connections—In wood construction, connections often limit strength; so many common failures of deck construction lead back to connection performance. Proper connections of deck joists to beams, beams to posts and decks to houses are critical.

Because fasteners and hardware in wood decks can corrode, it's prudent to minimize dependency on them. Wherever possible, joists and beams should bear directly on posts. This type of connection requires more vertical space, but it's more reliable than transferring load through fasteners.

There are a number of ways to connect beams to posts (drawing left). Two-by lumber can be used as a beam if either set directly on top of the post or let into a notched post. This notched connection only works when the posts are 6x6 or better because notching a 4x4 post with 2x side members leaves only ½ in. of post for you to bolt through.

A better option when supporting a built-up beam with a 4x4 post is nailing a ½-in. treated-wood spacer between the two 2xs and setting the beam directly on top of the post. You also can tie the connection together with a hot-dipped galvanized beam-to-post connector. Just remember, though, that whenever you cut notches or install lag screws or bolts in deck lumber—even if it's preservative-treated lumber—you should treat the openings in the lumber with a wood preservative.

The connection at the house must be detailed carefully—Attaching a deck to a house is risky business. Screwing or bolting into a house opens the siding's protective envelope to moisture, which can lead to decay and insect

Drawings: Vince Babak

attack. Wherever practical, it's best to build a freestanding deck.

If a freestanding deck isn't feasible, take extra care attaching the deck to the house. And although it probably goes without saying, nails aren't adequate to make this connection.

To prevent water from entering the house, it's important to caulk pilot holes in the band joist of the house before installing screws or bolts. It's also prudent to add spacers, such as a few washers, between the two structures to allow the gap between the deck and the house to dry. You also should extend metal flashing under the siding above the deck and over the siding below the deck (drawing p. 64).

If the deck is attached to the house, it may be necessary to reinforce the band joist of the house to resist lateral forces that tend to pull the deck from the house. Sixteen-d nails at 8 in. o. c. that are driven through the sole plates and mudsills, from above and from below, add support to the band joist in new construction.

It may also help to provide additional bracing on the deck; however, our recommendation to reinforce the band joist highlights the need to transfer adequately the deck loads to the house framing. There have been cases where the deck was firmly attached to the band joist, but the band joist was not secured to resist the deck loads and was ripped from the house when the deck failed. Of course, we recommend X-bracing between the posts for freestanding decks. This topic is covered in our deck manual in more detail (*Wood Decks: Materials, Construction, and Finishing*, published by The Forest Products Society; 608-231-1361).

Proper size and spacing of fasteners is critical—Wherever you use bolts in a deck, the strength of the connection depends on the correct size and spacing of the fasteners.

To attach decks to the band joist of a house, where you use 12-in., 16-in. or 24-in. joist spacing, two ⅜-in. dia. lag screws are needed every 24 in. for a 6-ft. span. Two ½-in. dia. lag screws are needed every 24 in. for spans of 6 ft. to 16 ft.

Don't skimp on fasteners—The two most important things to remember when choosing deck fasteners—framing nails, decking nails, screws, joist hangers, bolts and lags—are their holding capacity and their resistance to corrosion. Inadequate fasteners or improperly installed fasteners can cause connections to loosen, and when they corrode, they weaken the surrounding wood.

Most fasteners are made of mild steel or stainless steel and are produced in a variety of styles. Protective coatings are often applied to mild steel fasteners. Stainless-steel fasteners last the longest, followed by hot-dipped galvanized-steel

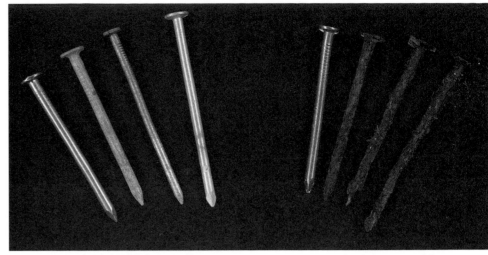

Stainless steel lasts longer. The nails on the right side of each pair were nailed into solid-wood blocks and subjected to 14 years of exposure to high humidity. From left, stainless steel, hot-dipped galvanized, mechanically galvanized and electro-plated galvanized.

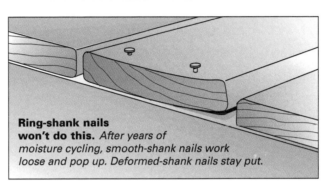

Ring-shank nails won't do this. *After years of moisture cycling, smooth-shank nails work loose and pop up. Deformed-shank nails stay put.*

fasteners. There are newer types of fastener coatings on the market, but we haven't extensively evaluated their longevity.

It's important to remember that aluminum fasteners can be used for fastening untreated wood but that aluminum can rapidly corrode in wood treated with preservatives containing copper.

Make sure your galvanized fasteners wear a heavy coat—Galvanized coatings protect the steel underneath, so when the coating is gone, the underlying steel corrodes. That makes the thickness of this protective coating critical.

To galvanize fasteners, manufacturers apply coatings of zinc, cadmium or zinc/cadmium by electroplating, mechanical plating, chemically treating or hot dipping (dunking the fastener in molten zinc). The thickness of these coatings varies significantly; hot-dipped coatings are typically the thickest and in our experience give the best corrosion resistance.

Unfortunately, many builders use electroplated nails for outdoor construction because they are available for use in nail guns. Our research found that electroplated nails don't last as long as hot-dipped galvanized nails (top photo).

In addition to nails, there are lots of hangers, post supports, hidden deck-board fasteners and other metal hardware available for use in deck construction. Just as with nails, screws and bolts,

metal deck hardware should have a thick, durable, protective coating.

Despite the cost, stainless steel is a bargain—Stainless-steel nails, bolts and screws can cost many times what conventional fasteners cost, but considering the overall investment of lumber and time put into a deck, they're worth the price, especially in wet or salty environments. Our research shows that even after years of severe exposure, stainless steel holds up well.

The problem with stainless steel is that the metal is softer and more difficult to drive than carbon steel, which may result in more waste from bent nails or damaged screw heads.

Avoid smooth-shank nails, and avoid nail pop-up—After years of getting wet and drying out, smooth-shank box and common nails can lose their withdrawal resistance, pop up and loosen connections, especially if they're used to secure deck boards (bottom photo). So for deck boards, we recommend deformed-shank nails, such as spiral-groove and ring-shank nails, or screws.

These deformed-shank nails resist withdrawal effects from cupping and from wetting-and-drying cycles. Pop-up can also occur when nails are too short. We recommend the use of at least 3-in. long nails (10d) to secure 1-in. thick deck boards and 3½-in. long nails (16d) for thicker deck boards.

Screws—especially drywall-type or bugle-head "multipurpose" screws—seem to have found a niche in deck building, too. Like other metal fasteners, screws used outside must be able to withstand the wetting-and-drying cycles that can cause weakening of metal and loosening of connections. Screws have advantages over nails: They are effective in drawing down cupped or twisted decking, and they can easily be removed. For screws, the length recommendations given previously apply. A word of warning about multipurpose screws, however: They are not intended to fasten joist hangers. Use only manufacturer-specified hanger nails to attach joist hangers.

Use lag screws where bolts can't go—For fastening a 2x to a thicker member where a through bolt won't work, lag screws work well. Just remember that pilot holes should be 60% to 70% of the diameter of the threaded portion of the screw. Therefore, a ⅜-in. dia. lag screw would get a ¼-in. pilot hole for the threaded portion, followed by a ⅜-in. pilot hole for the unthreaded portion.

Lag screws need to be long enough so that at least half of their length penetrates the thicker member. A flat washer should be used under the head, but not tightened so much that it crushes the wood.

Bolts are more rigid and typically stronger than lag screws. Just remember to drill the pilot hole

no more than ¹⁄₁₆ in. larger in dia. than the bolt. It's best to use flat washers under both the bolt head and the nut to distribute the force over a larger area and to reduce crushing of the wood.

It's also a good idea to saturate pilot holes with wood preservative or a water-repellent preservative (such as ISK Woodguard, Daps Woodlife or Cuprinol). Water can collect around fasteners and promote decay. Check lag screws and bolts periodically for tightness.

We haven't tested many of the newer fasteners, such as hidden fasteners, so we have no data. However, the use of hidden hardware may make it more difficult to replace a problem deck board should the need arise. On the plus side, these products don't puncture the top of the deck board with a fastener, eliminating a site for water collection.

After all of that time and money, give your deck a proper finish—A lot of time and money goes into building a deck. To keep it looking good and to ensure that it lasts, the deck needs a good finish. Unless you apply a finish, discoloration, checking and permanent damage can occur even with preservative-treated wood.

In general, wood finishes fall into two categories: those that form a film and don't penetrate the wood, and those that don't form a film and penetrate the wood. After a great deal of research, we recommend penetrating finishes (bottom photo, facing page).

Film-forming finishes include paints of all descriptions, solid-color stains, varnishes and lac-

quers. Penetrating finishes include solvent-borne, oil-based water repellents, water-repellent preservatives and oil-based semitransparent stains. Film-forming finishes usually lead to failure because the film can't tolerate the moisture cycling of deck lumber (top photo, facing page). Once the film is cracked, water gets under it, and the finish blisters and peels.

Choose a finish that really soaks in—Water repellents and water-repellent preservative pretreatments penetrate to protect wood. These products contain a moisture inhibitor, such as paraffin wax, and a binder, but not necessarily pigment. The amount of water repellent in the mixture varies among brands. A low concentration of repellent is about 1%, so it can be used as a pretreatment. Others have a high concentration of water repellent—about 3%—and are stand-alone finishes. If the label says "paintable," the finish probably contains the lower concentration of water repellent.

The difference between a water repellent and a water-repellent preservative is that the preservative contains a mildewcide. The use of a mildewcide even in a finish applied to preservative-treated wood is recommended because the wood preservative doesn't resist mildew.

Water-repellent preservatives also are available in forms that contain nondrying oil solvents such as paraffin oil. These products penetrate the wood but don't dry inside the wood.

Several commercial wood treaters are marketing 5/4-in. radius-edge decking that has a dual treatment of water repellent and copper chromated arsenate (CCA) preservative. This lumber is marketed under brand names such as UltrawoodR, Wolman ExtraR and Weathershield. Although this process is relatively new and its long-term performance isn't well-established, we believe these products are probably worth the extra cost.

Generally, dual treatments are used on #1 grade lumber rather than #2, which is a more common grade for treated lumber. Therefore, some of the increase in price reflects the use of this better-quality wood. We believe that the use of water repellents and water-repellent preservatives does increase the life of fasteners; however, we have never quantified this. We have found that these treatments can decrease iron staining if poor-quality fasteners are used.

Stain finishes are good if not overapplied—Semitransparent oil-based stain finishes penetrate wood, provide color and often contain water repellents or water-repellent preservatives. Some manufacturers make semitransparent "decking stains," which have enhanced water repellency and better wearing resistance. Don't confuse decking stains with siding stains, which

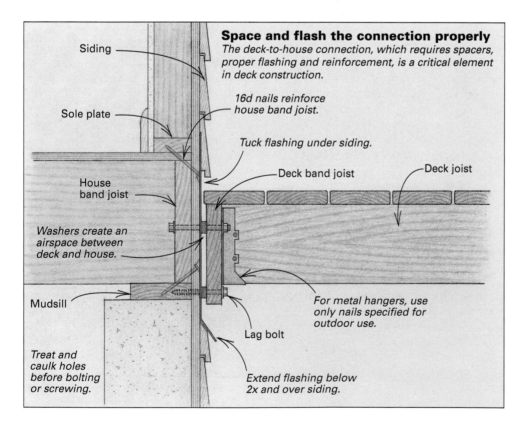

Space and flash the connection properly
The deck-to-house connection, which requires spacers, proper flashing and reinforcement, is a critical element in deck construction.

Siding

Sole plate

House band joist

Washers create an airspace between deck and house.

Mudsill

Treat and caulk holes before bolting or screwing.

16d nails reinforce house band joist.

Tuck flashing under siding.

Deck band joist

Deck joist

For metal hangers, use only nails specified for outdoor use.

Lag bolt

Extend flashing below 2x and over siding.

aren't for use on horizontal wearing surfaces. If you apply too many coats of stain, a film will form on the wood, and it eventually will crack and cause problems (photo center). If applied properly, semitransparent oil-based stains penetrate into the wood without forming a film.

Semitransparent deck stains last much longer than clear water-repellent preservatives because the pigment protects both the wood and the preservative from the damaging effects of the sun. One problem with stains is that the stain may wear off in high-traffic areas such as steps, and it may be difficult to hide these patterns completely when restaining.

Preservative-treated wood shouldn't affect the finish—Waterborne preservative treatments such as CCA don't affect the finishing characteristics of wood and may enhance the durability of some semitransparent stains. CCA contains chromium oxides that bond to the wood, decrease degradation of the surface and increase durability of semitransparent stains, often by a factor of two to three.

Other common wood preservatives don't contain chromium oxides, so staining this type of treated lumber is similar to staining untreated wood. Nonchromium treatments include ammoniacal copper zinc arsenate (ACZA) and ammoniacal copper quaternary (ACQ).

Don't put off applying the finish—On a newly built deck, apply the finish after the wood dries below about 20% moisture content. (For more on moisture content in deck lumber, see article on pp. 60-61). If your lumber is not preservative-treated and is grade-stamped S-DRY (surface dry), KD (kiln-dried) or MC-15 (average moisture content 15%) or is treated and stamped KDAT (kiln-dried after treatment), it can be finished immediately. If treated and stamped S-DRY, KD or MC-15, that only means it was dried before treatment. Ideally, these boards should be finished prior to installation so that the end grain of each board can be coated.

It's often recommended to wait a year to finish a deck. We think a year is too long to wait because checking, cracking and splintering can occur. We don't think you should wait more than two months to finish your deck.

Brushing on the finish is best, but follow the manufacturer's recommendations. You can apply the finish faster having one person spraying and another person following and working the finish into the wood with a brush.

To avoid lap marks in semitransparent stains, brush the stain on only two or three boards at a time and stain along their full length. Second coats of semitransparent stains should be applied while the first coat is still wet (within 30 minutes to 45 minutes), or they won't absorb. If

Film-forming finishes aren't good for decks. Paints, varnishes and other finishes that form solid films are bad for use on decks because of exposure to sunlight and moisture cycling.

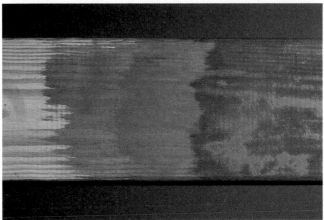

Too many coats have the opposite effect. More than one coat of semitransparent oil-based stain can be applied as long as subsequent coats are applied while the first is still wet and as long as not so much is applied that a film forms on the surface.

Let the finish soak in. The difference between film-forming and penetrating finishes is clear. The finish on the left is latex paint, which forms a film and isn't recommended for decks. Middle is penetrating water-repellent stain, and on the right is a penetrating water repellent; both are good on decks.

the first coat is dry, it seals the surface, and the second coat forms a film.

To maintain the water-repellent finish of your deck, it's best to reapply a finish annually or semiannually. The most obvious way to tell if your deck needs refinishing is to see if water beads on the surface or is absorbed. If water beads, there is no need to refinish. If it doesn't, apply a water repellent. If mildew is a problem, refinish with a water-repellent preservative. Usually, water repellents and water-repellent preservatives can be applied over existing finishes; however, it's always a good idea to test compatibility in an inconspicuous area.

If you refinish a deck finished with a semitransparent stain, be careful not to build up too much finish. Wait long enough that pigment loss is evident; or apply a clear water repellent or water-repellent preservative over the existing semitransparent stain for extra water repellency. □

Bob Falk, P. E., and Sam Williams are structural engineers at the USDA Forest Products Laboratory in Madison, Wisconsin. Their colleagues Andy Baker and Mark Knaebe contributed to this article. Falk and Williams are coauthors of a handbook on wood decks, Wood Decks: Materials, Construction, and Finishing.

A Deck Built to Last
Thoughtful details designed to beat the weather

by R. W. Missell

We get a lot of pleasure from the birds and other wildlife that live near our house here in Virginia. So when we decided to build a deck, it was important that it provide a transition from our living space to theirs. The house backs into the woods, and we wanted a graceful deck that wouldn't obstruct the view. On many other decks I have seen, the assembly of railings and supporting posts masks the view with a clumsy grid of horizontal and vertical lines. But I also wanted to make sure that the deck would stay strong and solid for decades, so it features a number of details that are designed to ward off the destructive effects of the weather.

The design stage—I usually know how the parts of a finished project should fit together. But as a novice carpenter, I had to learn some hard lessons about the nitty-gritty of putting the parts in place. I found that it's less frustrating (and less expensive) to work problems out with a pencil rather than with a saw.

To prepare for this project, I worked up a scale drawing that included a plan view, elevation views and details. I also drew—and drew again—lots of pencil sketches to help me understand how each piece would be made and how each joint would fit together.

I had to balance code requirements and aesthetics. For example, our county's policy requires 4x4 posts to support a deck railing. I knew these would look bulky, so my permit application included calculations showing my railing design (2x5 posts on 42-in. centers) to be stronger than required (4x4 posts on 60-in. centers). This data is found in tables entitled "Properties of Sections" in *The Wood Book* (Hatton—Brown Publishers, Inc., 610 S. McDonough St., Montgomery, Ala. 36197.

I wanted a deck pattern that would be decorative and that would visually break up the 30-ft. length of the deck. After looking at a variety of options, I decided to use a herringbone pattern made from 2x4 deck boards, run at a 45° angle. Making each "panel" 5 ft. wide meant that the longest deck board would be just over 7 ft. long, so I could start with standard 8-ft. boards. This minimized expensive waste.

Once I had the design squared away, it was easy to make up an accurate bill of materials. As a retail customer, I was able to negotiate only a relatively small discount on the initial bulk order. I knew I'd pay dearly for what I forgot—both in dollars and in lost time. On the other hand, liberal estimating could leave me with a lot of leftover materials that would be hard to store or to resell. In the end, my detailed material take-off was well worth the effort. On clean-up day, the only wood left over was a few treated 2x4s, which were quickly used in a landscaping project, and some 2x2s, which now serve as tomato stakes.

Figuring a foundation—With the planning complete and the lumber stacked in the garage, safely out of the sunlight to keep it from warping, construction could begin. One side of the deck would be supported by three separate posts. Two of the posts were supported on a concrete-block retaining wall, which had been built two years earlier with the future deck in mind. The rebar-reinforced blocks had been filled solid with concrete, and the footing was oversized in order to support the deck weight. When it came time for the deck, all I had to do was install anchor bolts for the posts (photo at right).

The third post had to be located in an existing sidewalk. I used a hammer drill fitted with a masonry bit to "perforate" a 12-in. square area of pavement, and then chipped the rest away with a hammer and chisel. Then I was able to excavate a pyramid-shaped cavity 24 in. deep, so that the finished footing would spread the load on undisturbed soil below the frost line. The hole was filled with concrete around wire-mesh reinforcement, and a J-bolt was inserted to anchor the post.

Sizing structural elements—A primary goal was to build a solid deck that would last at least 20 years and require an absolute minimum of maintenance. For this reason, I was conservative in sizing support members for strength and very careful at those points where wood rot could be expected. Although not required by our code, the galvanized steel bases I put under each post completely separate them from the concrete. A wood-to-concrete connection is one of the first places to rot, but the steel support provides a ½-in. air gap, thus ensuring good ventilation at the end grain of the post.

I decided that the posts would be 6x6s and the lengthwise support beam would be a pair of 2x12s. Local building code permitted 4x4 posts and a 2x8 beam, and upgrading these key components added only $55 to the cost of the project—a good long-term investment.

Where the support beam meets a post, the post was double-notched at the top in order to support a 2x12 on either side of a 1½-in. thick tongue. I sandwiched 2x12 scraps between the 2x12s of the support beam wherev-

The deck rests partially on a concrete-block wall that had been built some years before, but with the later construction of the deck in mind. The cores of the block had been reinforced and poured solid, so holes had to be hammer-drilled into the concrete for the anchor bolts (as shown in this photo), which were later grouted in place.

er a splice was required, and also at the ends of the beam. All of these components were through-bolted with ½-in. galvanized carriage bolts. The completed substructure formed a rigid, integral unit that helped to make the finished deck rock solid (photo adjacent page).

Ledger details—The support beam, located just beyond the lengthwise centerline of the deck, carries about two-thirds of the deck's weight. The ledger, however, is equally important from a structural standpoint because it ties the whole deck structure to the house as well as supports the remainder of the deck. Attaching the ledger board was a simple matter. I just removed the siding and used lag screws to attach the ledger board to the rim joist of the house.

The decking boards would have to be nailed to the ledger, and it was clear at the design stage that this wouldn't be easy. The 2x6 ledger provided only 1½ in. of nailing surface,

and even this was partially obstructed by the siding. As a result, the deck boards would have only an inch of bearing surface on the ledger. This connection would certainly be a weak link in the structure. And beside that, I didn't like the idea of driving all those nails so close to the aluminum siding, a material easily scarred by errant hammer blows. Instead, I nailed in a 2x4 flat to the top of the ledger (bottom photo, next page). The flat side of the 2x4 solidly supports the ends of the deck boards, and each of the 2x8 joists hanging from the ledger would have to be notched to clear the nailer. But the nailer was worth the trouble, because it reduced the likelihood of split end grain in the decking and also kept hammer dings in the siding to a minimum.

Once the siding had been removed to install the ledger, the house's main structural system was exposed to the elements. Rain and snow would reach into the voids and joints at the end of each deck board, and it

was vital to protect the untreated rim joist from this moisture. The solution was a simple flashing strip, which I slipped under the siding and extended over the top of the nailer strip to form a drip edge well away from the rim joist (photo bottom left). Admittedly, this flashing was perforated with nail holes when I nailed in the deck boards, but the joint between the nailer and the rim joist will still be protected. Next time around, though, I think I'd caulk beneath the flashing, too.

Joist details—For a given load, the correct joist size depends on how far they span and on how closely they're spaced. But figuring the proper spacing wasn't so easy for this deck. First, the diagonal pattern of the decking boards meant that each board would span a greater distance than the distance between joists. Second, a joist would be needed wherever the decking boards changed direction (every 5 ft.). I settled on a spacing of roughly 14 in. o. c. for the joists, resulting in a clear span of roughly 19 in. for the longest deck board. With this reduction in spacing, 2x8 joists were a conservative choice.

Along each line where decking boards changed direction, I doubled the joists, and spaced them apart with scraps of 2x. This provides a full 1½-in. support area at the end of each deck board, and nails can be sunk well away from the ends of the boards, which virtually eliminates end splits. It also provides an open space between joists so that a single pass with a circular saw trims all deck boards evenly. Most important, however, it ensures that the butt ends of the deck boards will be well-drained and ventilated. Tightly made joints directly over a joist will eventually rot as water collects around them. The doubled joists are supported by heavy-duty joist hangers (United Steel Products Co., 703 Rogers Drive, Montgomery, Minn. 56069).

Blending the design elements—All construction details to this point were designed to increase the strength and durability of the deck. But after the joists were in place, appearance became a major consideration. The deck's dominant visual element is the sanded and rounded 2x surfaces. All the 2x4 deck planks and exposed boards in the railing and

The key element of any deck is the support system of posts, beams and joists. In this project, posts were notched to support a beam built from doubled 2x12s. The post extends between the 2xs as a tongue and spaces them apart. Bench supports were solidly bolted to the substructure (above). Below, deck joists hang from a ledger lag-bolted to the house. The black joist hanger on the left is wide enough for doubled joists spaced apart with 2x blocking. Flashing slipped under the siding directs water away from the house. Just visible beneath the flashing is a 2x4, nailed flat to the top of the ledger that provides additional support to the deck boards.

The completed deck includes herringbone decking and a privacy screen at one end. Note the rounded front portion of the benches, a detail designed to increase comfort. Edges on the benches and on the railing were sanded smooth.

benches were belt-sanded smooth, and the edges were rounded over with a router and a ½-in. radius corner rounding bit. For assembly-line efficiency, I set up a separate work area for these operations, including a bench vise, sawhorse supports and power cords.

The railing system—Several conflicting design goals affected the construction details for the railing. First, I wanted the view from a seated position on the deck to be as unobstructed as possible. My intent was for people on the deck to feel themselves a part of the adjacent woodland, and not feel fenced out by the deck rail. The final design uses evenly spaced, slender horizontal and vertical lines to create a see-through effect, much like a venetian blind (photo above). A key factor here is the use of 2x5 posts turned edgewise to the deck to minimize their mass, and edgewise 2x stock for railing and barrier slats. The county safety code requires a 36-in. rail height with openings between railing and deck held to a maximum of 6 in., but the thin profile of the edgewise stock allowed me to meet this requirement with a minimum of visual impact.

Besides keeping a slim silhouette when viewed from a chair on the deck, the railing was designed to offer a broad, flat area for flower pots, serving plates and cold drinks. To achieve this, I used a 2x8 with rounded edges for the railing cap and a 1x apron for trim. Each joint in the rail cap was mitered across the thickness of the cap and carefully fitted to provide a smooth, unbroken surface.

The heart of the railing system is the 4½-in. wide strip of treated ½-in. plywood that ties all the pieces of the cap together (photo top right, facing page). Installation wasn't too difficult. After ripping the plywood to width, I screwed lengths of it to the tops of the posts. Then I cut the 2x8 rail caps to length and attached them to the plywood with 2-in. galvanized woodscrews run in from beneath

Installing benches single-handed can be tough, but the liberal use of clamps can speed the job. The bench backs and seats were angled slightly to make them more comfortable.

through the plywood. I ripped the trim pieces from 1x stock and rounded the edges before nailing them to the underside of the railing. As a result, no nails are visible from above, and no plugs are required in the rail. Equally important, the plywood acts as a splice plate to keep butt joints tight and to reduce warping and cupping of the rail cap boards.

Bench building—I cut the bench supports from 2x8 stock, wanting to make sure they would be strongest where each support is bolted to the deck frame. I didn't want them to look clumsy, so each support is tapered to a dimension of 1½ in. by 4½ in. where it joins the railing (photo above). The bench supports were installed at an angle of 15° so that the seat back would be comfortable. The seat rails

were cut from 2x6s and mounted with 5° of tilt. Combined with the rolled edge at the front of each bench, these angles result in very comfortable seating. I used galvanized ½-in. carriage bolts to tie everything together.

Privacy screen—In order to block the view from a neighbor's yard, I added a privacy screen at one end of the deck (photo top). The screen was built in the same way as the rail system, but with more horizontal slats—these were spaced ½ in. apart. I added an extra support near the diagonal section of railing in order to support the ends of the slats.   □

R. W. Missell is an electrical engineer with 20 years of experience in the operation and maintenance of Air Force facilities.

A Quality Deck

Built-in benches, recessed lights and a spacious spa highlight a high-class deck

by John Baldwin

Some time ago, Chris and Judy Lavin hired our small design/build firm to craft a fancy cherry wet bar and remodel the kitchen in their 1950's home. The completed work enhanced the home's interior, but the exterior was another matter. For one thing, its T&G cypress siding was cupped, checked and shedding paint. For another, the existing doors and metal casement windows had suffered from decades of New England weather. Equally significant, the ample backyard was unmanicured and unused.

Based on the quality of our earlier work, we were commissioned to replace the existing cypress siding with oiled 1x6 V-groove Western red cedar, and the old windows and doors with new wood ones. The heart of the job, though, would be in the backyard. That's where the Lavins wanted us to build a 1,200-sq. ft. hot-tub deck. At least, that's what they called it. When we looked over the drawings supplied by architect John Gardner Coffin, we discovered that the so-called "deck" would be a lot more like a piece of built-in furniture than an exterior add-on (photo right).

The plan—Attached to the north side of the house, the deck was to be accessible from the kitchen and living room through a series of sliding-glass doors. Instead of the usual handrail, it would be enclosed by a continuous 21½-in. high bench. This substitution was acceptable to the local building department because the bench would be relatively wide (about 18 in.) and because the surface of the deck would be relatively low to the ground (5-ft. maximum above grade).

The bench would be clad on both sides with vertical 1x6 cedar siding, which would mesh visually with the newly refurbished exterior of the house. Enhancing this effect, the cedar on the outboard side of the bench would extend below the rim of the deck to within about 2 in. of grade, forming a skirt that would conceal the deck framing and concrete piers.

Coffin also called for a set of stairs to fan out from the deck into the newly sodded backyard. A second stair, adjacent and parallel to the house, would allow easy access to the east end of the yard.

After discussing the project at length, the Lavins decided to equip the benches and adjacent house soffits with recessed low-voltage lighting so that the deck could be used at night. They also settled on an 88-in. square Sundance "Cameo" spa (Sundance Spas, 13951 Monte Vista Ave., Chino, Calif. 91710; 714-627-7670) as the deck's centerpiece, which at the time was the biggest four-seater on the market (photo, p. 73). Finally, they requested that a built-in planter be incorporated into the perimeter of the deck, and that several more planters be built on wheels so that they could be moved around.

For longevity, Coffin specified that the deck be supported by poured concrete piers and the bottoms of the two stairs be supported by concrete-block foundations, with the piers and foundations extending below the frostline (42 inches in this part of New England) to prevent frost heave. He also called for the use of pressure-treated Southern yellow pine for the framing and decking (though we amended the decking material later) and the use of galvanized nails and metal connectors throughout.

Trenches and craters—Even before we broke ground, we encountered our first snag: as designed, the new deck would cover the existing septic tank. The Connecticut Wetlands Department (whose domain seems to include any land that ever gets wet) required a minimum clearance of 10 ft. between the deck and the tank. This meant that at considerable expense, we would have to bypass the old septic tank and install a new tank and leach field the prescribed distance from the deck. We were also required to excavate a 1,000-gal. drywell for the spa. We informed Chris that "we can build your deck, but your backyard's gotta go." For-

tunately, as a former builder, he understood Murphy's Law and gave us a quick go-ahead. The project started in late June, the beginning of the hottest summer Connecticut had seen in 30 years.

With the septic system rearranged, we set batter boards and ran string grids to plot the centers for the 24 poured-concrete piers. We used powdered lime to lay out the locations of the two buried 8-in. concrete-block walls that would support the stair stringers. The holes and trenches were dug with a backhoe.

After forming the 8-in. by 16-in. concrete footings that would support the block foundations, we shoveled 1½-in. to 2-in. crushed stone into all the trenches and pier holes, forming level pads 42 in. below grade. The tops of the piers would extend 8 in. above grade, so we formed them with Sonotubes—cylindrical cardboard tubes available from masonry suppliers.

Before pouring the concrete, we determined the finished elevations of the piers using a transit level, and then cut each Sonotube about 10-in. short. During the pour, we rested the Sonotubes on the gravel pads, filled them

roughly one third full of concrete, and then pulled them straight up 10 inches. This let a good measure of concrete slump out the bottoms of the tubes so the footings and shafts of the piers would be cast in a solid block.

Once the Sonotubes were filled with concrete, we set a ½-in. by 12-in. anchor bolt in the center of each one and troweled the tops smooth. The anchor bolts allowed us to bolt a TECO galvanized post anchor (TECO Products, P. O. Box 203, Colliers, W. Va. 26035; 800-438-8326) to each pier. These anchors have a cam-lock design that allows more than 1 in. of play in positioning the deck posts, while at the same time raising the posts about 1¼ in. above the tops of the piers to prevent the posts from wicking moisture from the concrete. Once we laid up the block walls and backfilled around the piers and walls, we were ready to start framing.

Basic framing—The pressure-treated framing for the deck consists of 4x4 posts and built-up girders (doubled and tripled 2x8s) supporting 2x8 joists spaced 16 in. o. c. For

Clad with Philippine mahogany and Western red cedar and washed by incandescent light, the deck is more a piece of built-in furniture than a simple platform.

drainage, we sized the posts so that the deck would slope away from the house ¼ inch per 4 feet. At the house, the joists are supported by a 2x8 ledger fastened with galvanized lag bolts and lead shields to the existing concrete-block foundation.

Because its base would be positioned 16 inches below the surface of the deck, the spa needed a platform of its own. We built a 7-ft. by 10-ft. platform that also allowed access via a hinged desktop hatch to a control panel mounted in the side of the spa (more on that later). Supported by concrete piers and 4x4 posts, the platform was framed with 2x8 joists spaced 16 in. o. c., then topped with 1¼-in. by 5½-in. pressure-treated pine. This was sufficient to support the more than 5,000 lb. the spa would weigh when full of water and hot-tubbers. A short knee wall built around the pe-

rimeter of the spa platform supports the edge of the deck above.

The stair stringers consist of 2x10s that rest on 2x6 pressure-treated plates anchor-bolted to their supporting concrete-block foundations. We laid out the stairs so that each riser is the height of a single deck board (5½ in.) and each tread is three deck boards deep. To reinforce the longer stringers adjacent to the house, we nailed a 2x4 to the side of each one.

The bench chassis—The bench framing consists of two short 2x4 frame walls—an inner one and an outer one—capped by a third one laid flat (drawing, p. 73). We aligned the outer walls so that the outside faces are plumb with that of a continuous, horizontal 4x4 beam lag-screwed to the outsides of the deck posts. This allowed the outer wall and the beam to serve as a nailing base for the vertical cedar siding. Of course, positioning the outer walls of the bench this way placed them outside the rim joists of the deck instead of over them. To support the outer walls, we nailed a 2x4 ledger to the outboard sides of the rim joists.

Also, where necessary we shimmed the 4x4 beam straight.

Where the bench runs perpendicular to the deck joists, its inner walls are secured with 16d nails to the joists. Where the bench parallels the joists, its inner walls are supported by 2x8 blocking installed 16 in. o. c. between the joists. To ensure that the bench would be as straight as possible, we cut its framing components accurately by clamping a stop to the fence of our miter saw.

We left one section of the bench open temporarily. That allowed seven of us to slide the spa from the flatbed trailer when it was delivered and lower it onto its platform. The spa was then plumbed and wired before the decking was installed.

Switching to mahogany—With the extensive deck framing nearly completed, everyone was tired of coping with pressure-treated Southern yellow pine sawdust and splinters, yet we still had to install 1,200 bd. ft. of decking. One of our carpenters had recently built door jambs out of Philippine mahogany, and he suggested that we use mahogany instead of pressure-treated pine for the decking.

Because Chris had been designing, building and racing power boats for years, he knew that the durable, rot-resistant wood was commonly used for boat decks and hulls. He also liked its appearance—the color of Philippine mahogany ranges from deep maroon to almost pure white, and its grain figure from quarter-sawn to bird's-eye. Southington Specialty Wood Company of Southington, Connecticut, offered us the mahogany at an attractive price, and because they were nearby, their shipping costs were reasonable. To everyone's delight, Chris opted for the mahogany, although it added about 10% to the cost of the deck.

Before we started on the decking, we secured a length of ¾-in. plywood to the bed of our 10-in. miter saw. This plywood raised the 1¼-in. by 5½-in. deck boards far enough off the miter-saw table that the sawblade could cut through the stock in one pass. To help support the ends of the decking, we extended the plywood about 2 ft. past either end of the saw, leveling the plywood wings with wood blocks. Finally, we nailed a narrow wood strip on top of the plywood about 5 in. to either side of the sawblade, perpendicular to the fence. These fulcrums raised the ends of the deck boards just high enough to produce a 2° back bevel for a tight fit at the butt joints.

We soaked the end grain with CWF wood preservative (The Flood Co., P. O. Box 399, Hudson, Ohio 44236; 800-321-3444) before installation. Originally, we had planned to screw and peg the decking to the joists, but decided to use 12d spiral-cut galvanized nails instead to cut costs. The mahogany took nails extremely well; whenever a board split it was invariably due to a preexisting check. To avoid damaging the wood with hammer marks ("elephant tracks"), each of us knocked a small knot out of a shim shingle so that we could slip the shingle over the nails before driving

Lighting the basement. The basement was originally lit by casement windows housed inside a concrete areaway. To prevent the deck from dimming the basement, removable mahogany grates (photo above) were installed directly over the areaway.

The access hatch. The spa's control panel (photo left) is reached by way of an access hatch attached to the deck with a stainless-steel piano hinge.

Planting the perimeter. At the client's request, a planter (photo below) was built into the deck's perimeter. It's enclosed by a bench topped with mitered mahogany decking.

Bench framing

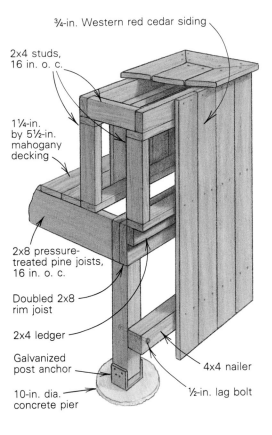

- ¾-in. Western red cedar siding
- 2x4 studs, 16 in. o. c.
- 1¼-in. by 5½-in. mahogany decking
- 2x8 pressure-treated pine joists, 16 in. o. c.
- Doubled 2x8 rim joist
- 2x4 ledger
- Galvanized post anchor
- 10-in. dia. concrete pier
- 4x4 nailer
- ½-in. lag bolt

The Sundance "Cameo" spa weighs more than 5,000 lb. when full of water and hot-tubbers. It's supported by its own platform, 16 in. below the surface of the deck.

them home (after all, nobody's perfect).

The 25-in. by 45-in. access hatch next to the spa (bottom left photo, facing page) consists of 2x6 pressure-treated pine joists spaced 16 in. o. c. and topped with the mahogany decking. The heavy hatch was originally designed to be lifted up out of its opening, but this would have invited back problems. Our solution was to hinge the hatch to the deck with a 45-in. long stainless-steel piano hinge fastened with stainless-steel screws.

Making the grates—Before the deck was built, Chris's basement workshop was illuminated by two casement windows housed inside a 2½-ft. wide by 12-ft. long concrete areaway (a sunken space that allows air and light into a basement) adjacent to the house. To prevent the deck from blocking this light, we installed three removable 2½-ft. by 4-ft. mahogany grates directly over the areaway (top photo, facing page).

The grates consist of 1¼-in. wide by 1⁹⁄₁₆-in. thick mahogany strips ripped out of leftover deck boards, lap-jointed into a grid-like pattern and supported by 2x4 ledgers. We cut the lap joints using a table saw equipped with a dado blade and a 40-in. long wood fence C-clamped to the saw's standard miter gauge. A small wood guide pin notched into the bottom of the wood fence was positioned so that after each lap joint was cut, the workpiece was repositioned with the newly cut joint stopped by the pin the correct distance from the next cut. This ensured perfect, uniformly spaced lap joints. We nailed the grates together with 6d galvanized finishing nails, then used a router

to round over all 408 openings in the grates. The finished grates let plenty of light into the basement, and each is sufficiently rigid to support several hundred pounds.

Mitering the bench tops—The plans called for the use of three parallel deck boards for each segment of the bench top, with the boards crosscut at each end. For a more elegant look, we crosscut the center boards only, capping the end grain of each one with a short length of mahogany mitered to the two outside boards (bottom right photo, facing page).

To prevent splinters and to soften the bench visually, we rounded over the ends of each center board using a router fitted with a ⅜-in. roundover bit. We laid out the miters in place by lapping the two boards to be mitered, marking their intersection at the inside and outside edges, and then scoring the cut line with a straight edge and a utility knife (where the benches met at a 45° angle, the miters were more than 11 in. long). We cut the miters with a circular saw and fine-tuned the joints with a smoothing plane, giving each miter a shallow back bevel for a tight fit at the top. Before nailing down the bench tops, we installed 2x10 pressure-treated blocking beneath the location of the mitered joints and smoothed the top of the bench framing with a power plane.

Lighting it up—Before the benches were boxed in with cedar siding, they were wired to accommodate several lighting fixtures. Once the siding was applied, we cut small rectangular openings in it to house the fixtures.

Coffin called for the use of Prescolite model

37G-1 lighting fixtures (Prescolite, Inc., 1251 Doolittle Dr., San Leandro, Calif. 94577; 415-562-3500). These fixtures each hold a pair of 25-watt incandescent light bulbs concealed behind a plastic louvered grille that sheds weather and directs the light downward toward the mahogany decking. Combined with recessed lighting in the house soffits, the bench lights provide plenty of glare-free illumination for socializing.

Finishing off—In addition to a fixed 4-ft. by 4-ft. planter (bottom right photo, facing page), we built three rolling planters for the deck. They're basically 2-ft. cubes made of pressure-treated frames, weather-proof casters, cedar siding and mahogany trim.

Our final job was to apply CWF wood preservative to the deck and planters. We applied two coats to the edges of the decking using a 1-gal. pressurized plastic garden sprayer with a narrow nozzle. A paint roller fitted with a 10-ft. extension handle was used to finish the surface of the deck. Bench tops were rubbed with progressively finer grades of steel wool between multiple applications of preservative. The residue was soaked up with a tack cloth.

Much to the clients' delight, the overall impact of the deck is striking. By day, the variegated color of the mahogany plays off the oiled cedar. By night, when the blue light from the spa constrasts with the raking yellow glow of the built-in lighting, the effect is almost dreamlike. □

John Baldwin is a carpenter and writer in Greenwich, Conn. Photos by Bruce Greenlaw.

A Sound Deck

This wrap-around deck roots house to hillside overlooking Long Island Sound

by Elizabeth Martin

The site is magnificent, a rare find in suburban New York. Overlooking Long Island Sound, it affords a distant view of open water, a closer view of islands, and closer still, a view of a wildlife refuge on a spit of land maintained by the state. At low tide, the inlet at the property's edge becomes a sand flat—an ideal habitat for shore birds. On his second visit to the site, architect Stephen Tilly watched a deer swim from the wildlife refuge to one of the islands.

The house on the property had started as an unobtrusive cottage perched among trees on a bluff. But former owners had added a garage, a children's wing, a greenhouse and an enclosed pool. Concerned only with the view from the inside looking out, they had ended any possibility that the house would nestle inconspicuously in the trees. It had become an ungainly assemblage.

Tilly's client was thinking of buying the property and wanted to know if the interior of the house could be improved and the living spaces extended outdoors to take full advantage of the site. Only then would he consider the high price the location commanded. After inspecting the site and the plans, Tilly concluded that interior

Elizabeth Martin is a landscape designer who works with Stephen Tilly in Dobbs Ferry, N. Y.

changes would be no problem. Adding on outside, however, would be tricky business.

The client wanted a deck overlooking the water that was on the same level as the swimming pool and living quarters. According to Tilly, the proposed deck "would be sitting out there in space, a long way above the steep rock hillside. This would mean long columns that might be pretty ugly. The terrain looked hard to work on and tricky to get footings into. Not only that, the landscape is so nice you don't want to have stuff coming down into it." Former owners had apparently tried to mask the massive underpinnings of their house with Japanese landscaping.

A crucial plus for the project was the successful bid of Branch Woodworking to do the job. Trained in a preservation workshop of the National Historic Trust and formerly employed in the reconstruction of period rooms in the American Wing of the Metropolitan Museum of Art, contractor/cabinetmaker Ben Branch made an exception to his usual rule of avoiding deck and siding jobs when he took on this project. The design and site stimulated him. And for him, his partner (and brother) Chucker Branch and carpenter Gary Bromley, working where they had a view of waterfowl and sailboats seemed a fine way to spend the fall.

Design—The client liked California living and wanted the easy flow of indoor/outdoor activity in his own house. He wanted the new deck to have the refinement of an indoor space and at the same time share qualities with the surrounding landscape. The deck had to be accessible to interior living areas for daily use, and also large enough to accommodate lots of guests.

To meet these requirements, the deck evolved into a band skirting the southern and eastern faces of the house. Tilly made the deck relatively narrow outside the main living areas so as not to obscure their view of the Sound (drawing, facing page). After all, for much of the year the client would be seeing the water from indoors. This concern, together with the constraints imposed by the steep site and the beautiful rock formations midway along the eastern foundation led him to place the widest part of the deck diagonally off the pool area.

Before the deck was added, the eastern elevation of the house began abruptly with a concrete foundation rising unrelieved for 12 ft. This severe foundation wall is the first in a series of vertical lines marking additions to the house. The lines slice down the eastern facade, making the building seem higher than it is and divorced from the landscape. As Tilly saw it, one of his jobs was to try retroactively to root the building in its environment in a more sensitive way.

Tilly designed a sitting area three steps (21 in.) below the main deck at the southeast corner. Here built-in benches provide modest enclosure and serve as a resting place in the treetops out of deck traffic. Viewed from the Sound, the lowered "outdoor room" helps step the house down the hillside. From inside the house and from the main deck, the water is fully visible beyond the bench enclosure.

Still fighting the vertical segmentation, Tilly avoided running the deck outline parallel in plan to the building outline. Instead he modulated the deck width, creating passageways and gathering places of various sizes.

"Scale was a key issue," says Tilly. "If I was going to make any impact on the bulwark, it was going to have to be up to scale. The house is an overpowering mass; deck elements had to be big to matter."

Standing off the northern end of the deck, 8x8 posts form an oversized gateway-arch. Derived from the traditional *torii* of Japanese shrines, it frames the first view of the Sound for people coming around the north end of the house. The gateway-arch extends the deck into the landscape, and the beefier posts relate to tree-trunk dimensions. "I thought of the house as a cliff. Then beside it I built a redwood forest—the deck posts—which then would go into the existing forest on the hillside."

To punctuate changes in the deck outline, Tilly extended key beams and joists as much as 3½ ft. beyond the floor. The curved ends of these structural members afford a still smaller scale, this one relating directly to people on the deck.

To keep the view open, the railing for the main part of the deck needed to be as transparent as possible. It was built with one 2x4—its narrow dimension facing the house—as the handrail and three strands of ¼-in. steel cable (photo facing page).

Pricey materials—The original specs called for a combination of redwood and pressure-treated lumber, with cedar as an alternative. But after Branch priced the range of materials, the client elected to use redwood on the entire deck. Branch had his supplier commit to a price for the entire job, which was important because the cost of redwood fluctuates so much. Later they had to buy individual pieces at other yards, and these cost up to three times as much as the pieces in their original order (part of the difference, of course, reflects the economy of bulk ordering). Some boards cost well over $3/bd. ft., so individual pieces, a 20-ft. 2x10 for example, could cost nearly $100.

Stainless-steel cable for the railing and related hardware was also expensive and hard to locate. A single galvanized eye-bolt cost over $3, a lag bolt, $6.50. A local marine-supply store, where they got the galvanized hardware, finally procured the cable for them.

Branch wasn't sure how much better galvanized hardware is than zinc-plated. Calls to suppliers during the ordering process never fully convinced him that the tenfold difference in price was justified. And despite the expense and

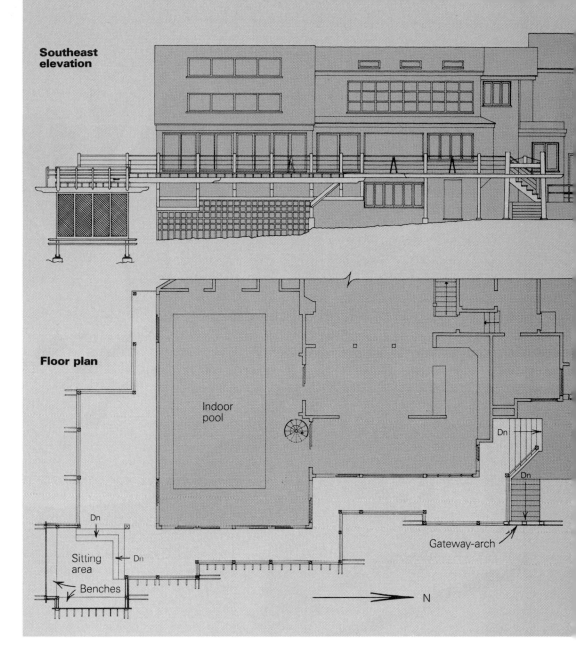

Southeast elevation

Floor plan

Indoor pool

Dn

Dn

Dn

Sitting area

Dn

Benches

Gateway-arch

N

fastidiousness in avoiding oxidizing metals on the project, stains are beginning to show under the holes where the stainless-steel cable passes through the railing posts. Branch speculates that iron is at the core of the stainless braid.

Setting up—One of the few accommodating things about the site from the construction standpoint was a roadway, apparently left over from an earlier construction project, that led up to the side of the pool enclosure. The builders stacked several truckloads of lumber at the end of the road, close to the beginning of the deck. Then layout work commenced.

It helped that previous work on the house was well done: plumb and square, with level sills and clapboards. The existing building could be used as a reference, and moreover, the new work would jibe with the old. Tilly had keyed locations for some of the new posts and braces off existing features, like the mullions of the pool enclosure, to integrate old with new and to minimize obstruction of the view. This meant precise layout in relation to the existing building, and consequently, some uneven spacing in the new work.

Several large boulders sat right where footings were required. These were a struggle to remove

without having them crash down through the landscaping to the ocean. With the boulders out of the way, the builders could then dig footing holes in the rocky hillside. Most of the 2-ft. square holes reached 36 in. to 40 in. down.

In some of the steep locations, the builders hit sloping ledge near the surface. Worried that the footings might slide into the Sound at some point, they jack-hammered holes in the rock and inserted rebar to tie the footing to the rock below.

After piling aggregate, sand and cement near their lumber at the road's end, the Branches began to load their rented half-yard concrete mixer. They used a wheelbarrow to get the concrete to the first two footings, but the terrain forced them to hand-carry it in 5-gal. buckets to the next 14 holes.

The Branches were careful not to spill on the landscaping or exposed rocks and did their best to minimize the visual intrusion of the footings on the landscape by integrating them into rock formations and strategically moving boulders around later (top right photo, next page).

Meanwhile, ledgers were going up around the pool enclosure. There were two ledgers here, an upper one for the joists and a lower one for the diagonal braces. The builders hammer-drilled ¾-in. holes for lead shields every 18 in. and

used lag bolts to attach the ledgers to the reinforced concrete-block pool foundation.

Where the ledgers attached to wood-frame sections of the house, they bolted through the band joist to get a good grip instead of just using lag bolts. But rather than drill blindly, they opened up the basement ceiling to see if their drills would hit anything important and to examine the framing members. The Branches also removed clapboards above the ledgers so that they could install flashing.

The Branches were satisfied that the masonry anchors were fastened about as well as they could be into concrete block. The block had not crumbled, the holes had been drilled cleanly and cleared of dust, and the shields seemed snug. But Tilly points out, "The Branches flagged a problem in the field that my consulting engineer and I had not picked up on the drawings. Ben questioned how the part of the structure near the pool would resist the outward thrust from the diagonal braces. In the middle of the pool wall we would be pulling out on the masonry anchors holding the top ledger to the block wall rather than pushing in on them. The whole assembly was still pinned at each end, but repeated heavy loading at the edge of the deck could start to loosen the middle of the ledger."

After a visit to the site and a review of the original drawings for the pool addition, Tilly and the engineer decided to bolt the ledger through the block wall, using some conveniently located openings for air-conditioning ducts in the slab. Generous reinforcing meant that the slab and the block wall were tied together and would act as a unit. Chucker eyeballed routes, sometimes angled to avoid obstacles, and drilled holes for ½-in. threaded rods 18 in. long. These were fastened through ¼-in. steel plates, which acted like huge washers to spread the stress in the duct openings. Three air-conditioning cavities in the slabs received two each of these rods—securing the ledger in a total of six places. The result allowed everyone—builders, architect and engineer—to sleep comfortably at night.

Framing—Footings and ledgers secured, the builders started setting posts. They bolted down adjustable post bases over anchor bolts that had been embedded in the concrete, and raised the still uncut 6x6 posts one by one and tied them temporarily to the ledger with 2x4 bracing.

On the recommendation of their supplier, the Branches had originally ordered construction-heart redwood posts. But these were unavailable, and so even the posts were clear all-heart redwood. Everyone agreed that the extra cost was worth it, especially with a design that fea-

tured posts so prominently, but it meant the builders had to use galvanized nails even for temporary braces to avoid any bleeding on the precious redwood.

The longest post was nearly 20 ft. and took three men to maneuver it on the steep slope. They roughly plumbed and braced the posts, and then made fine adjustments when they brought the framing out to embrace the posts.

On the Sound side, the deck is supported by 6x6 diagonal braces bolted between pairs of 2x8 joists (bottom photo, facing page). The Branches made up the braces on the ground, propped up pairs of joists on temporary legs, dropped in the 6x6 diagonal brace and fastened everything together. These assemblies were then braced side to side, and the Branches walked precut 6x6 railing posts out to the end of the braces and lowered them in between the 2x8 joists. Blocks tacked to each side of the posts stopped them at the desired height, and the builders plumbed and bolted the posts in place.

A pair of 2x10s, running underneath the ends of the joists, connects the post legs. "Our original intention," says Ben Branch, "was to assemble things on the ground and then install them. But it seemed that even with three or four guys, a board over 30 ft. long was just too flimsy. It handled like spaghetti." So they cut scarf joints in the 2x10s at strategic locations and connected them in place with marine epoxy (WEST System, Gougeon Brothers, P.O. Box X908, Bay City, Mich. 48707) and bolts.

Decking, railing and detailing—The builders put off laying the deck boards as long as they could, for fear of damaging them, and walked around on plywood. But as framing progressed, Gary Bromley began to lay the 5/4 x 4-in. redwood, using 8d galvanized nails.

Once the deck was down, everyone worked to complete the detailing. While Ben built benches, Chucker and Gary worked on railings. The railing system involved putting tension on three ¼-in. stainless steel cables strung between

the posts (photo p. 74). The cables were resisted in compression by the bullnosed 2x4 top rail, which was toenailed side and bottom to the posts with 16d galvanized finish nails.

Initially they used a come-along to pull the cables through the galvanized eye-bolts screwed into the posts, but they soon discovered that they could do it by hand. They borrowed a special tool (Nicopress, made by National Telephone Supply Co., 5100 Superior Ave., Cleveland, Ohio 44103) from the marine-supply store to crimp the metal sleeves that formed the loops at the end of the cable.

Gary cut joist and beam ends, following a template supplied by the architect. The steps at the northern end of the deck are angled on one side so as not to block the path down to the water (photo above). Short stringers butt into the side of a header stringer that cuts across the plan view at a 45° angle. The stair railing was stiffened with galvanized pipe let in underneath.

The builders used another pipe to brace the gateway-arch back to the deck. The 1-in. copper pipe encloses a ¾-in. threaded rod fastened through the posts at each end. It resists the tension in the rod and prevents bowing.

When the deck was nearly finished, everyone agreed that the longer railing spans needed stiffening. The railings were great for the view, but a little disturbing to lean on. Tilly designed an inverted V of ½-in. square aluminum rods, which were then made by an aluminum fabricator. The builders installed them midspan in all rail sections 6 ft. or longer. They're held in place with finish nails at the bottom, and the apex of the V sits in a hole drilled into the bottom of the rail.

Including a short interruption to batten down and weather out Hurricane Gloria, the deck took two months to build. The client and his family are delighted with their new location. He feels that having the Sound as an extension of his home is a renewing influence in his life. Every morning at dawn he jogs in the wildlife refuge. As he looks back on the shore, he can see his house growing into its site on the bluff. □

On the north end of the house, the angled steps leave room on the terrace to enjoy the view of Long Island Sound framed by the ceremonial gateway-arch, visible at the far left.

Opening the doors doubles the room. A generous, 6-ft. wide door opening coupled with matching floor levels creates an intimate connection between the garden deck and the dining room. Cedar trim inside and out intensifies the bond.

A Dining Deck

A deck that's built to withstand the weather is united with a small dining room

by Tony Simmonds

O ne of the joys of living in Vancouver, British Columbia, is that when the weather is warm enough, you can open your house to the outdoors without being eaten alive by insects. So it's a natural feature of local design to include a deck or some outdoor living space in your house plans. I enjoy the challenge of integrating outdoor spaces with the rooms that border them.

Bill Abbott and Kris Sivertz's house had a bad deck in a good location. Right off a small eating area in the kitchen (photo facing page), the deck was on the south side of the house between the kitchen and the backyard pool. That was the good news. Unfortunately, the deck was detailed poorly, and as a consequence it was falling apart from rot. It was also one step down from the floor level in the house. Besides being a potential hazard, this broke the continuity of floor surface, which is one of the essentials of good indoor-outdoor connections.

The other essential is a generous opening between the two, and the 5-ft. patio slider Bill and Kris had didn't qualify. Even the widest of sliders fails in this respect because half the opening always is obstructed. What you need is hinged French doors; by opening them up wide, you truly can abolish the barrier between inside and outside.

Here, though, we had a problem. There wasn't space on either side of the opening for a 2-ft. 6-in. door, let alone the 3-ft. doors I planned to use. They were going to have to be bifolds, an arrangement certain to cause some difficulties. I consulted Peter Fenger, custom sash and door man, and he didn't say it couldn't be done, so we plunged ahead.

Benches and planters take the place of railings—To feel like an extension of the indoors, an outdoor space needs some enclosure. Usually, walls are suggested by a railing and the ceiling by a trellis. But the 42-in. guardrail required by code can make a small deck feel like a crib. Luck was on our side in this instance, however. Because the deck is less than 2 ft. above grade, we were free from the railing-code constraints. So I planned benches for two sides of the deck, enclosed and supported at their ends by planters (top photo).

I believe strongly that an indoor/outdoor space needs at least the suggestion of a ceiling. So I wanted an open trellis of 4x4 beams above the deck. But I had to work hard to convince Bill and Kris of the wisdom of this idea. Given the scarcity

Partly in, partly out. Built-in benches and planters at the corners make a boundary around the sides of the deck, and the spare trellis overhead implies a ceiling. The planters brace the posts at the corners.

Inset straps align the doors. Cocobolo straps fit into grooves in the door frames. When the doors are folded, the straps are retracted. Note how the doors are slightly different widths, which allows the doors to fold flat to one another without the doorknob getting in the way.

of sun that shines here, people are reluctant to put anything between themselves and the sun.

Deck details for a wet climate—Bill and I framed the deck with pressure-treated 2x8 hemlock joists, supported on a ledger at the wall and on a triple 2x8 built-up beam. Instead of spiking the 2x8s together, I used the code alternative of ½-in. dia. bolts, 40-in. o. c. I used extra washers to space the 2x8s about ½ in. apart—enough to allow air to circulate as well as to make the outer faces of the beam flush with its 6x6 posts. Making the beam the same width as the post allowed us to attach the beam and the post easily with steel connectors instead of toenails.

Before setting the beam on them, however, I covered the top of each post with a piece of EPDM roofing membrane. I also capped the beam with 2x8 blocks between the joists (top drawing, p. 81). The blocks are sloped to divert water and dirt off the top of the beam. The blocks also provide secure nailing for the joists. In the spots where I needed doubled joists, I used pressure-treated plywood spacers between them to promote drainage and ventilation. I think precautions such as these are cheap insurance. The fewer paths there are for moisture to get into the framing, the better.

The planters are structural—At first glance, the cross section of the planter boxes might look like a 2x wall sheathed on the inside with treated plywood and on the outside with T&G siding (bottom right drawing, p. 81). But they are plywood boxes covered with framing and ing. The plywood contributes the rigidity; the framing provides a flange for securing the 4x4 posts affixed at the three outer corners of each planter to support the trellis.

Bill and I assembled the boxes with Fastap self-drilling screws (13909 NW Third Court, Vancouver, Wash. 98685; 800-874-4714). These screws are coated with a nongalvanized proprietary finish called Duracoat that is supposed to be more durable than galvanizing. More importantly, the coating isn't corroded either by the copper in pressure-treated wood or by the tanins in red cedar.

To overcome the problem of trying to get the tops of four posts to make a perfectly straight line where they meet the beam, we installed the two outer posts, then the beam, then the two intermediate posts.

Like the posts, the beams are clear, pressure-treated 4x4 cedar. We anchored the beams to the posts with 8-in. galvanized helix nails (you predrill for these babies). Where the beams intersect each other, the upper beam is notched out to a depth of 1 in. Counterbored screws secure the connection, and the counterbores are filled with teak plugs installed with clear silicone instead of glue.

Next, build the benches—The benches are suspended between the planters atop an egg-crate frame of interlocking 2x4s, bolted through the planter walls at the ends and supported in the middle by a 3x6 post (photo p. 80). The post is screwed from underneath to a 3x6 beam, which also bears on the planter framing. To ac-

Keep the water out of the framing. An interlocking frame of 2x4s supports the built-in benches. At midspan, a short 3x6 post supports the frame. The top of the post is capped with metal flashing to shed water, and the bolted supports affixed to the planters are shimmed out with washers to create an airspace.

commodate the flashing (bottom left drawing, facing page), this beam had to be installed before the siding went on the planters. But the eggcrate frame was bolted on after the siding was in place. Once again, I used a half-dozen extra washers to maintain airspace between the frame and the siding.

The 2x4 seat slats for the benches were installed with hidden fasteners called Dec-Klips instead of nails (Ben Manufacturing Inc., P. O. Box 51107, Seattle, Wash. 98115; 206-776-5340). Dec-Klips are a good way to avoid nailing through the face of the decking, and one of these days I'll use them for a whole deck.

The backs of the benches make a good show of being joinery but actually are put together entirely with screws. I considered using biscuit joinery and waterproof glue, but I shied away from that for two reasons. First, it would have taken longer to do and would have required kiln-dried stock; and second, I didn't trust rigid glue joints to hold up under the kind of flex and stress I knew the back would take.

I made up the backs as ladders around which I assembled the frames. The back slats are 5/4 KD red cedar. All other parts are also clear red cedar but milled from green 2x4 stock selected for dryness as much as for grain pattern and straightness. Toward the end of summer, you can sometimes find some pretty dry wood in the piles out in the yard, especially shorts, which don't sell as fast. I used Fastap screws for these assemblies, drilled and counterbored, and plugged where exposed with teak plugs that I left proud and sanded lightly. (Unless you're going to seal plugs in with paint or varnish, you might as well leave them proud deliberately because if you cut them flush, they'll pop out when they expand.)

The 12-in. Delta portable planer I bought not long ago got almost as much use as the radial-arm saw on this job. Thicknesses graduate throughout the seat backs, from the full 2x top rail to the 15/16-in. finished thickness of the 5/4 slats. Shadowlines and stepped joints are a practical and beautiful Greene & Greene legacy for which I thank them daily. The top rail, for example, started out as a 2x4. I planed it smooth, then ripped it to just less than 3-in. The offcut, about 5/8-

in. thick, was planed to 1⅛-in. wide and trimmed to the length between the 1¼-in. thick main uprights. It became the means of fastening slats. The completed seat backs are joined to the benches with screws from underneath and to trellis posts through notched wings at either end of the top rail.

The planter caps are sloped to drain—The planter caps are 2x6s tapered in section (bottom right drawing, facing page). I milled the taper by running the 2x6s through the planer while atop a sloped board. Before planing the slope, I ran drip kerfs in the bottoms of the cap material ½ in. from each edge.

The caps were mitered using biscuit joinery for alignment, and they were fitted tightly around the trellis posts. Then I used a straightedge and razor knife to cut a channel about 3/16-in. square through the joint, which I filled with a marine-grade polysulfide caulk, Sikaflex 231 (Sika Corp., 22211 Telegraph Road, Southfield, Mich. 48034; 800-967-7452). This caulk comes in the usual three or four colors. I used black—not subtle, but handsome to my mind, and reminiscent of boat decks. This is a time-consuming and finicky detail, but so far (I've used it on only one other deck, built in the spring of 1991) it appears to be successful in eliminating Curling Miter Syndrome, which is so painful and familiar to outdoor woodworkers.

Planters, benches and the simple lattice between the posts were finished with two coats of clear Duck'sback Total Wood Finish (Masterchem Industries, P. O. Box 368, Barnhart, Mo. 63012; 800-325-3552), an exterior finish that goes on milky and dries perfectly clear without appreciably darkening the cedar. The trellis and decking are pressure-treated and were left unfinished. Although you can use Duck'sback on treated lumber, too, it's a good idea to let the treatment be absorbed and to let the wood dry for at least two months before application.

All the doors fold outward—Early mornings and late nights during the building of the deck had been given to head-scratching about those bifolding French doors. I had to draw something

for our doormaker, Peter Fenger, so he'd know the job was for real and plug it into his schedule. But I changed just about every detail before he built the doors. By that time, when Pete saw me coming through the door, he would get that look on his face. "This is it, Tony," he finally said one morning. "We're cutting."

Here's what we came up with, and why (bottom photo, p. 79). First, the doors had to open onto the deck; the dining room was too small to lose floor area to them. The fold would bring the inside faces of each pair of doors together. To allow them to fold flat without the handle getting in the way, the two center panels are wider than the outside ones. Serendipitously, this also makes a nicer rhythm of proportion than doors of equal width, I think.

Four doors meant eight stiles taking up glass area, so we didn't want the stiles to be any wider than necessary. At the same time, a sealed unit made with laminated glass on both sides is not light. We settled for 3-in. wide stiles, with a 4½-in. top rail and a 7-in. bottom rail. To add strength and to make room for a ½-in. airspace between the panes, we made the sash 2¼-in. thick. The additional thickness also allowed us to use 4x4 hinges and still have plenty of wood for a rabbeted astragal at each meeting stile.

The biggest problem was closure hardware. How to pull four doors tight against their jambs without having handles at the folding hinge point? Large surface bolts were an option, but I thought it would be cumbersome to have to lock each door independently, top and bottom. Besides, even though the entry door in the adjacent wall meant the hardware didn't have to be designed for constant use, I wanted it to be clear what a person had to do to open the doors. A profusion of bolts seemed likely to confuse. Dummy handles on the inactive leaf of any pair of French doors contradict this principle, too, by offering an option that turns out to be false. How many times do we need to suffer that small embarrassment? If there is only one handle to grasp, on the other hand, there can be no confusion.

One handle, large enough to grasp comfortably, could be provided only by a cremone bolt because the door stile isn't wide enough for any

Photo this page: Tony Simmonds

Section through bench and planter

The planters do more than house plants. At their corners, they anchor the posts that hold aloft the trellis. On the inside, they support the benches. Note how the planter caps are sloped to direct runoff into the planters.

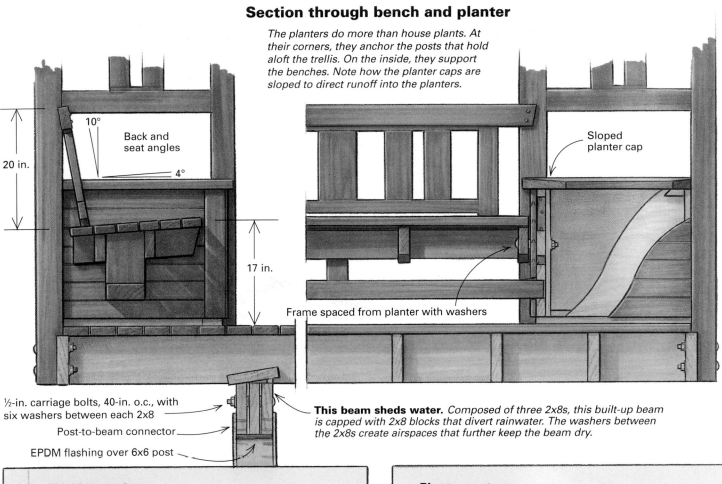

10°

Back and seat angles

4°

20 in.

17 in.

Frame spaced from planter with washers

Sloped planter cap

½-in. carriage bolts, 40-in. o.c., with six washers between each 2x8

Post-to-beam connector

EPDM flashing over 6x6 post

This beam sheds water. *Composed of three 2x8s, this built-up beam is capped with 2x8 blocks that divert rainwater. The washers between the 2x8s create airspaces that further keep the beam dry.*

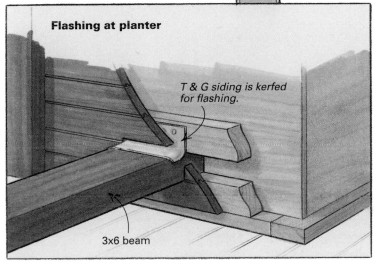

Flashing at planter

T & G siding is kerfed for flashing.

3x6 beam

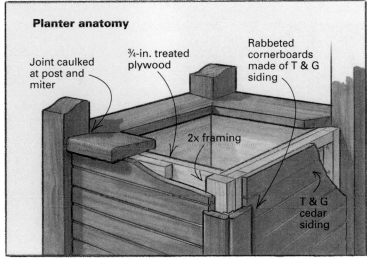

Planter anatomy

Joint caulked at post and miter

¾-in. treated plywood

Rabbeted cornerboards made of T & G siding

2x framing

T & G cedar siding

---

standard lockset. The principle of uncluttered surfaces and transparency of function suggested ordinary flush bolts in the edge of the inactive leaf. This combination meant the need of finding a way to lock each pair of doors into a single panel so that each could be opened and closed as a single door. I didn't know of any commercially available hardware capable of doing this, so I designed some wooden strap bolts for the job (bottom photo, p. 79). They fit into grooves cut into the rails and stiles at the tops and bottoms of the doors.

Time to cut the grooves—Plowing four sizable grooves in a set of $1,800 custom-made doors seemed a pretty earnest commitment to an untried design when it came time to do it, and I will admit to a degree of nervous procrastination

over the job. The night before, I made Bill look at mock-ups screwed to the doors where the bolts were to go to help me determine the exact length and width. As always, he was patient, optimistic and gently biased toward optimum mechanical efficiency. We settled on a width of 1⅛ in. and a length of 13¼ in., the maximum we could get without quite intersecting the line of the primary hinge stile.

I also wanted to make sure the doors were fitted to their openings before cutting the grooves for the bolts, so I installed spring bronze weatherstripping (Pemko Industries, P. O. Box 3780, Ventura, Calif. 93006; 805-642-2600) at the head and side jambs and between each door. An aluminum and vinyl door bottom (also from Pemko Industries) was fitted in the groove I had asked Peter to machine. Before installing the weather-

stripping, I sealed with oil all areas that couldn't be reached after the weatherstrip was in place.

I made the bolts and their retaining straps from cocobolo, a dense, hard wood that grows in Mexico. Each pair of doors was removed and laid on sawhorses to rout the slots, but the final fitting of bolts had to be done with doors in place. Achieving the right degree of resistance meant balancing the pressure of the weatherstripping by easing the inside surfaces of the retaining straps with sandpaper. The beauty of surface mounting retaining straps with countersunk brass screws is that if the bolts cause problems, it will be simple to make adjustments. □

Tony Simmonds operates Domus, a design/build firm in Vancouver, British Columbia. Photos by Charles Miller except where noted.

Curved Concrete Deck

Layout tricks and precise formwork create a sinuous addition to a square house

A deck in four layers. Cantilevered concrete decks encompassing a hot tub make the transition between a curved sunroom and a concrete patio on ground level.

by Greg Torchio

By most standards, the 4,200-sq. ft. house in suburban Washington, D. C., was already big. But it seemed almost modest when compared with some of its 8,000-sq. ft. neighbors. The Colonial-style house, essentially a big brick box, was built in the 1960s as a spec house and lacked some features that buyers expect these days. And because of small windows and a compartmentalized floor plan, the house was dark. When it was sold a few years ago, the new owners decided to make some changes, inside and out, that would open up the house and let in more light.

Their plans included the addition of a sunroom and the expansion of an existing concrete deck at the back of the house. That side of the house was three stories high, counting a walk-out basement, and faced southwest, so the orientation could accommodate a sunroom. The owners wanted the addition built off the main floor level, with the deck expanded so that it could include a hot tub. The new deck had to be durable, but the owners hoped it could be made of something other than treated wood.

I suggested that by using the same materials as the house, primarily brick and painted wood trim, I could make the addition fit the general look of the house without mimicking its boxy lines. If permanence was what they wanted in a new deck, concrete would be a good choice of materials. I thought brick, block and concrete also would be well suited for a more organic, fluid design that would soften the rectangular shape of their house (bottom photo, p. 84).

Final plans included curved brick walls for foundations and for the walls of the sunroom addition and a series of multilevel cantilevered concrete decks connected by sculpted concrete stairs. The second deck level, 12 ft. in dia., wraps around a hot tub (photo above). The free-flowing curves outlining the decks, the stairs and the sunroom allow various spaces to be defined without being separated. A curving cedar fence makes the hot-tub deck and the adjacent patio feel like private, outdoor rooms.

Among the major design constraints were local property setback requirements; the higher the elevation, the larger the required setback from property lines. This requirement led to our multideck approach, which allows lower-level decks to extend farther away from the house, providing plenty of outdoor living area but still meeting local codes. The decks also create a stepped transition between the main-level deck and the ground level. To make the lines of both decks and stairs merge smoothly and fluidly, I used the radii of 15 different circles in drawing the plans. The curves lock together to form a sinuous concrete and brick shape that adds a completely different flavor to the back of the house (drawing facing page). For all of its visual appeal, though, the plan also presented a challenge for my crew and me when it came to the layout and construction of the concrete slabs.

Foundations and deck-slab forms—Before any of the cantilevered decks could be poured,

foundation walls were built of 4-in. brick over 6-in. block (photo facing page). The walls beneath the main-level deck incorporated existing columns and support both the concrete slab and the walls of the sunroom addition above it. Built on conventional concrete footings, the addition's foundation walls enclose a utility/storage area. Foundation walls were also built for the lower hot-tub deck to provide an insulated enclosure for the hot tub's filter, pump and heater. Remaining areas behind foundation walls near the hot tub were backfilled.

After footings and foundation walls came the 6-in. thick deck slabs, and the first step was putting up a temporary support structure that could carry the weight of all that fresh concrete. Once the concrete hardened, the supports could be removed. To start, my crew and I set 6-ft. high pipe scaffolding on wood mudsills. The scaffolding was topped with adjustable twist jacks that supported 8-in. steel I-beams set at right angles to the house (top right photo, p. 84). On top of the steel, we placed 7-in. aluminum I-beams perpendicular to the steel I-beams 2 ft. o. c. Finally, ¾-in. CDX plywood was tacked down to wood flanges on the tops of the aluminum beams. This created a solid deck that ran far enough beyond the proposed slab edges to give the crew a place to work. This procedure was repeated for the lower elevated deck but without the scaffolding. Instead, twist jacks and timber cribbing were used to support the steel I-beams. In the low areas beneath the hot tub,

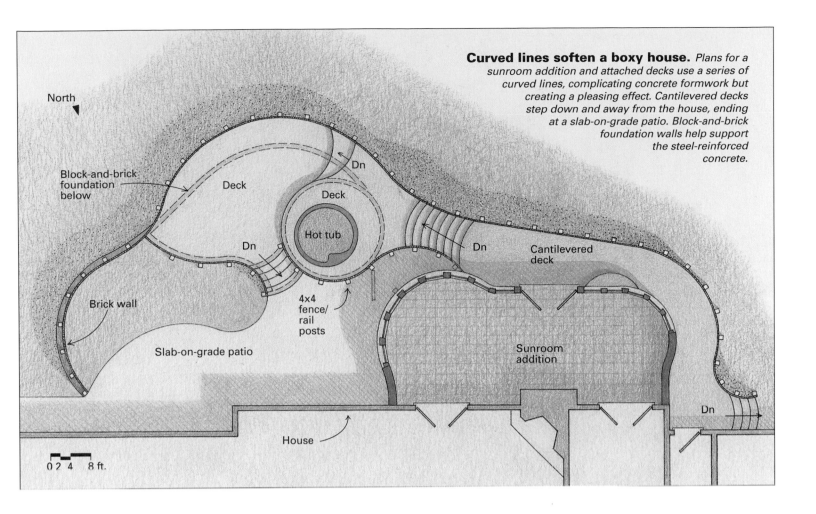

Curved lines soften a boxy house. *Plans for a sunroom addition and attached decks use a series of curved lines, complicating concrete formwork but creating a pleasing effect. Cantilevered decks step down and away from the house, ending at a slab-on-grade patio. Block-and-brick foundation walls help support the steel-reinforced concrete.*

North

Block-and-brick foundation below

Deck

Deck

Dn

Hot tub

Dn

Dn

Cantilevered deck

Brick wall

4x4 fence/ rail posts

Slab-on-grade patio

Sunroom addition

Dn

House

0 2 4 8 ft.

there was no room for the aluminum joists, so we used 4x4s instead.

Laying out curved slab edges—On the drawing board I had calculated the radii for all of the curved edges of the decks and located these lines relative to the existing building. Unlike layout lines for foundations, which could be laid out at grade, the centerpoints of the many curves for the elevated slabs had to be located at the correct elevation above grade. This was accomplished by various means. Some centerpoints, for curves at lower levels, were merely on high stakes. One radius for the upper deck was swung from a centerpoint located on a scaffold erected solely for that purpose. Other arcs were traced from big, curved templates cut out of plywood.

The lines of the deck edges were scribed directly on the plywood forms. The edges were formed by 6-in. high 2x4 walls. Curved plates for the walls and the hot-tub form were cut from ¾-in. plywood, and the walls were then sheathed with whatever thickness of plywood could make the bend—one layer of ½-in. plywood or two thicknesses of ¼-in. ply (top left photo, p. 84).

The new concrete deck was supported by and anchored to the existing slab on the main level with 6-in. by 4-in. ⁵⁄₁₆-in. steel angle. Steel studs ½-in. dia. and 6 in. long were welded to the steel 2 ft. o. c. to key into the new concrete. The steel was bolted to the existing slab with ½-in. dia. by 5½-in. long Hilti Kwik Bolts (Hilti Corp., P. O. Box 21148, Tulsa, Okla. 74121; 918-252-6000). The same

Bricks to match the house, curves to enliven it. Cantilevered concrete decks are supported by block foundation walls faced with brick to match the house. The circular foundation will support a hot tub and provide an insulated enclosure for the tub's filter, pump and heater.

Curved formwork. Snaking concrete forms were made from ¾-in. plywood plates cut with a jigsaw, 2x4s and plywood walls bent to the right shape. The circular form in the center of the deck created a cutout for the hot tub that was installed later.

Scaffold and I-beams support deck forms. Elevated concrete decks were poured over plywood forms supported from below by a cribwork of aluminum and steel I-beams. The 8-in. steel beams rested on adjustable jacks on the scaffolding, making precise leveling possible.

Extra steel to support cantilever. The last step before pouring the concrete was placing grids of #4 steel rebar and a PVC waterstop. The waterstop prevents leaks between the deck and the sunroom.

Forms stripped. Concrete decks, which are 6 in. thick and reinforced with steel, cantilever well beyond the foundation walls. Anchor bolts placed in the edges of the deck will be used to attach posts for a fence.

type of steel angle was used at the existing brick-and-block wall of the house and anchored with Hilti ½-in. by 8-in. HIT Renovation Anchors 1 ft. o. c. The steel was positioned with the long leg up and the short leg at the bottom to create a shelf that would support the concrete deck. The angle steel sat flush on the plywood form.

Ready for the pour—With all deck forms in place and covered with a form-release agent, it was time to place the steel rebar. The steel had been precut, bent, labeled and keyed to a shop drawing, but it took hours to sort through it all after it was delivered. The engineer-designed structural slabs included two grids of #4 rebar (½ in. dia.) 1 ft. o. c. held in place by high and low chairs (bottom left photo, above), which are small metal supports that hold the rebar during the pour. Little did I know that the nuisance of picking autumn leaves out of the steel every morning was just a prelude to the major role Mother Nature would have in this production.

Building in the winter in the Washington, D. C., area usually is not that big of a problem. The crew was finally ready for the first pour in December. As it turned out, that's when we got a winter's worth of snow. To guard against freezing, I ordered concrete with an acceleration additive for faster setup, and I increased the mix to

Sunroom addition. Walls of the sunroom addition also are curved, helping to relieve the visual monotony of the rectangular brick Colonial. The multilevel concrete decks are connected by poured concrete steps with curved treads and risers.

3,500 psi, 500 psi more than our engineer said we needed. The faster setup and increased strength limited the amount of time the concrete had to be protected from freezing and allowed us to strip forms sooner. It was interesting to see later that the seven-day strength of the concrete tested at or near 100% of its design strength.

I also invested in 10 concrete-curing blankets. The 6-ft. by 24-ft. blankets are made of ½-in. closed-cell polyurethane insulation covered on both sides with reinforced polyethylene. The covering has a sewn binding with grommets around the entire perimeter. Blankets help keep in the warmth and the moisture generated by curing concrete. Because large portions of the slabs were elevated and exposed to freezing from below, my final precaution was to tent the entire job with 6-mil polyethylene. Inside the tent I placed two kerosene torpedo heaters equipped with thermostats and was able to maintain a 40° to 55° air temperature. Even with all of that, I lost a little sleep thinking about the work that had gone into the forming, the reinforcing and the pouring and how all of it would have to be redone if the concrete froze.

Separate pours—The concrete was placed in three separate pours. The first was for the two main-level decks—the one at the main-floor level

Bottom photo, this page: Jefferson Kolle

and the one immediately below it where the hot tub would go. The second was for three sets of stairs and the curb on which the sunroom addition would be built. The last pour was a 2-in. finish slab in the sunroom and the lower patio slab on grade.

The structural slabs were straightforward except for getting the concrete to its final resting place. Some of the concrete was pumped, some was sent down a long site-built chute, and the rest was wheeled. This is the part of the job that we carpenters occasionally make the mistake of doing, either because the finishers we hire refuse to do all of the placing, or we just forget how hard it is. For one of the pours, I recruited Rick Goldstein from my architectural staff to help in the wheelbarrow parade. Now he too knows how losing a wheelbarrow full of concrete lifts the morale of the rest of the crew when it's most needed.

Where stairs would connect the levels, rebar dowels stick out of the decks, and a keyway was formed. Pairs of painted anchor bolts for the fence posts were placed in the slab perimeter about 4 ft. o. c. (bottom right photo, facing page). Dowels also were placed where the curb for the sunroom wall was going. Because the main deck and the curb for the addition walls were poured at two different times, I was afraid moisture would wick through the cold joint between them. To prevent that, I used a Horn PVC waterstop (AC Horn, Inc., 7405 Production Dr., Mentor, Ohio 44060; 800-218-2667). This product comes in various widths and profiles. The one I used was 6 in. wide by about ¼ in. thick. It was positioned upright with the bottom 3 in. embedded in the slab and the top 3 in. in the curb (bottom left photo, facing page).

The 6-in. by 6-in. concrete curb was formed in the same manner as the curved edges of the slabs—curved plywood top and bottom plates, 2x4s and plywood sidewalls. The main-deck slab has both heated and unheated spaces above and is unheated below. Because of this, 4 in. of extruded polystyrene insulation was placed on top of the slab in the area that would become the sunroom. Then a 2-in. slab was poured over the insulation to the height of the curb.

Pouring curved stairs—There are four sets of concrete stairs with curved sides and risers. At a minimum, they are 6 in. thick. To build the forms for the stairways, we used ¾-in. plywood supported from below by 2x6s 16 in. o. c. and 4x4 posts to the ground. The 2x6s and the plywood followed the slope of the underside of the stairs. The side forms were then built as curved walls bearing on the sloping floors. The riser forms spanned the space between the side forms and were made of ½-in. plywood sprung around the curved edges of 2x material placed on the flat (photo above). Four layers of plywood were added to the riser form to create a 2-in. deep by 5-in. high nosing (drawing above).

A word of caution when using concrete accelerators: Even in cold weather, the concrete sets up before you think it should. Forms for these concrete stairs included pieces to form the sides, the bottom and the riser—but nothing to contain

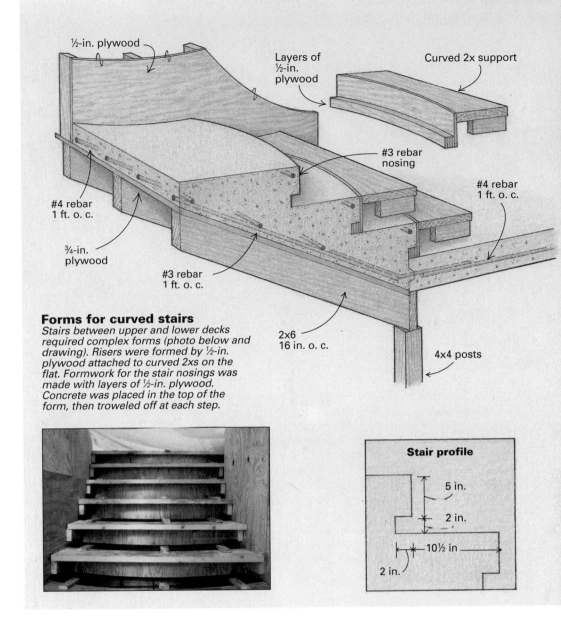

½-in. plywood

Layers of ½-in. plywood

Curved 2x support

#3 rebar nosing

#4 rebar 1 ft. o. c.

#4 rebar 1 ft. o. c.

¾-in. plywood

#3 rebar 1 ft. o. c.

2x6 16 in. o. c.

4x4 posts

Forms for curved stairs
Stairs between upper and lower decks required complex forms (photo below and drawing). Risers were formed by ½-in. plywood attached to curved 2xs on the flat. Formwork for the stair nosings was made with layers of ½-in. plywood. Concrete was placed in the top of the form, then troweled off at each step.

Stair profile

5 in.

2 in.

10½ in

2 in.

2 in.

the concrete for the tops of the treads. A common problem is that the concrete wants to rise out of the lower steps as the weight of the concrete above increases. The usual solution is to remove excess concrete with a trowel after it sets up a little so that the tread can be smoothed flat. Because of the additive, though, the stairs set up before we could get the high spots down. We had to chip off the high spots later and patch them. Because these decks got a top coating later, we knew our patching wouldn't be visible. The best patching compound I have found for this situation and for patching low spots on slabs is Euco Poly-patch (The Euclid Chemical Co., 19218 Redwood Road, Cleveland, Ohio 44110; 800-321-7628). For the deck, we chose a fiberglass-reinforced acrylic coating called All Deck (Environmental Coating Systems, Inc., 3321 S. Susan St., Santa Ana, Calif. 92704; 714-641-1340). We picked the product because it claimed to be durable and easy to clean. We found it to be neither, and I would not recommend All Deck.

Building the deck rail and the fence—The deck railing and the fence are important elements. The structure is quite an involved carpentry project, with an open rail on the upper deck and a closed privacy fence on the lower levels. It consists of vertical cedar 4x4s, each bolt-ed to two ½-in. by 1-ft. steel anchor bolts embedded in the edge of the concrete slabs. Where the fence sits atop a short brick wall, the 4x4s are bolted to steel stirrups embedded in the wall.

Two laminations of 1x4 cedar make up the curved rails. The first layer was bent into place and screwed to the posts; the second layer of cedar was glued with resorcinol glue and clamped to the first layer. This procedure was very time-consuming, and it required great patience, to say nothing of the many clamps we needed. My crew—Mark Holmes, Mike Mehalic and Jay Evans—got a lot of practice with both and did a great job of making the finished product look like bent 2x4s.

The rails were followed by the pickets. Like the posts, these were installed to a rough length only. Then, using a long, flexible strip of clear pine, we marked the curving lines at the top and bottom of the fence and cut the fence to shape. Top corners of the 4x4 posts were then rounded with a long blade in a jigsaw and sanded smooth. The entire fence and rail was painted to match the trim of the house and the addition. □

Greg Torchio is an architect and builder in Centreville, Md., and a one-time partner in Jersey Devil, a design-build group. Photos by Rick Goldstein except where noted.

The Deck Upstairs

How to combine a deck with a roof that won't leak

by Dan Rockhill

My first rooftop terrace introduced me to the built-up roofing business the hard way. I'd designed a small addition with a nearly flat roof, and I wanted to put a deck on top of it to take advantage of the summer weather, the view and the privacy. I called up the roofing contractors to find out how low the bids would go, and started dreaming of deck furniture.

As the roofers looked over the job, it became obvious that details and prices fluctuated wildly. One wanted insulation on top of the roof sheathing, another underneath; some wanted gravel in the flood coat, and others said gravel was unnecessary. To make matters worse, none of them would guarantee the roof because I wanted to cover part of it with a deck.

I finally got the job done, lost money and learned a lot. Since then I've done a lot of successful and more profitable installations, but I've never stopped keeping a keen eye out for good roof-terrace detailing.

Early decks—Years ago rooftop decks were called promenades or plazas, and were usually found atop fancy commercial buildings. They were covered with a hard and durable surface like quarry tile, laid over a built-up roofing membrane that kept out the water. This type of decking was occasionally used in high-end residential work, but it was too expensive for most ordinary construction.

For a more economical solution, designers looked once again to commercial work, and they came up with the system generally used today—walkboards over a built-up roof. Walkboards, which are simply planks nailed to supports called sleepers, were originally used for scuttleways to reach roof-mounted mechanical equipment on factories, warehouses, office buildings and the like. The sleepers spread the traffic loads, and kept the fragile roofing plies from being crushed.

There are two types of roofing membranes suitable for supporting walkboards or rooftop decks: the tried-and-true bitumen (asphalt or coal-tar pitch) built-up roof, and a modern alternative, the elastomer roof. The built-up roof, with its many layers of hot-mopped felt, is perfect for use on low-pitched roofs because it forms a continuous membrane that keeps out wind-driven rain.

Elastomers are synthetic polymers with elastic or rubberlike qualities. They have gained wide acceptance in commercial work, despite their higher price tag, because they offer excel-

lent flexibility, and resistance to weathering, fire, airborne chemicals, ultraviolet radiation and abrasions. They are also available in a variety of colors. Silicone, neoprene, Hypalon, acrylic and polyurethane rubber are all varieties of elastomers. Some are available in sheet form. Most can be applied as a liquid and finished with a trowel or paint roller. Toxic fumes can be a problem with these products, and you should be sure to pay special attention to instructions involving joining, mixing, curing times and using solvents. Because elastomers are still relatively new to the residential market, many roofers prefer to work with the traditional hot asphalt and gravel, and they continue to install the proven built-up roofs. What I will describe here is how to put a deck over a built-up roof so that the roof won't leak.

The substrate—A rooftop deck has to sit on a structure designed to carry the same loads as the interior floors of the house. Consequently, the members supporting the roof deck should be at least as sturdy as the floor joists below. To divert water away from the building and to prevent ponding, the roof should slope at least ¼ in. per ft. To get this slope, I usually nail tapers ripped from 2x4s to the tops of the rafters, but if the rafters are deeper than required for their spans, it's just as easy to cut the tapers on the rafters themselves.

The roof sheathing is no place to skimp on materials. The deck sleepers could wind up being placed between the rafters, and the sheathing will have to carry this load without flexing. I use ¾-in. CDX plywood over rafters 16 in. o.c., held down with construction adhesive and 8d cement-coated nails. I nail a line of blocks between the rafters to catch the edges of the plywood. For maximum strength, I stagger the joints.

Insulation should go between the rafters, and not between the sheathing and the membrane, where it would gradually get compressed from the deck loads, losing its effectiveness and causing the membrane itself to stretch.

Cants and base—Once the roof sheathing is nailed down, it's time to put on the cant strips and the membrane base. Cant strips are used wherever the plane of the roof meets a vertical surface. They provide a gentle 45° slope for the roofing felts to conform to. Be sure to use them. Felts forced to bend at 90° have a weak inside corner, which will fail before the rest of the membrane. Roofing contractors use fiber cants,

or you can rip and stack strips from 2x4s, as shown in the drawing on the facing page, bottom left. The legs of the triangle of a section of cant strip should be 3½ in. long.

Hot-mopping—You can install your own hot-mop membrane by renting a kettle and buying your supplies from a wholesaler, but I don't recommend doing this. Lifting pails filled with 525°F asphalt is a hairy job, nasty and dangerous enough to discourage even the most ambitious do-it-yourselfer. One wiggle of the bucket could spill the smoking asphalt and blister your skin or ruin the finish siding below.

I hire a subcontractor to do the membrane work. This isn't always easy either. Because it costs as much for the roofer to fire up the kettle for one square (100 sq. ft. of roof coverage) as it does for a hundred-square warehouse, many roofers aren't interested in a small job. So specify precisely what you want, then solicit bids. To make sure there are no surprises, don't accept any bids from contractors who don't look at the job in person. At current prices you can expect the roofer's share of a small deck job to be between $400 and $1,000. I also get an agreement before work begins about damage compensation in case a misplaced bucket of hot tar damages part of the building or grounds. Here's how things should go.

Base felt is a heavy, bitumen-impregnated sheet that weighs about 40 lb. per square and comes in 36-in. wide rolls. It is used only for the first layer, and it should be nailed down, not hot-mopped. Nailing it to the sheathing allows some movement between the substrate and the membrane, and makes it a lot easier years down the road to remove the roof if it should need replacing. It has to be firmly attached though, or a stiff wind will put it in the neighbor's yard. Nail it down with roofing nails spaced 6 in. o.c. on the edges and laps, and staggered at 18 in. o.c. in the field.

Membrane—Next, the base is covered with the membrane. It's made up of alternating layers of felt and molten bitumen. The felt acts as a reinforcement, while the bitumen bonds the layers of felt together and forms a waterproof film between each.

Each layer of felt in the membrane sandwich is called a ply. These felts are typically 15 lb. per square, and are held together by a matrix of fiberglass, asbestos or organic fibers. I prefer fiberglass felts because they have terrific

Structure of the roof and deck

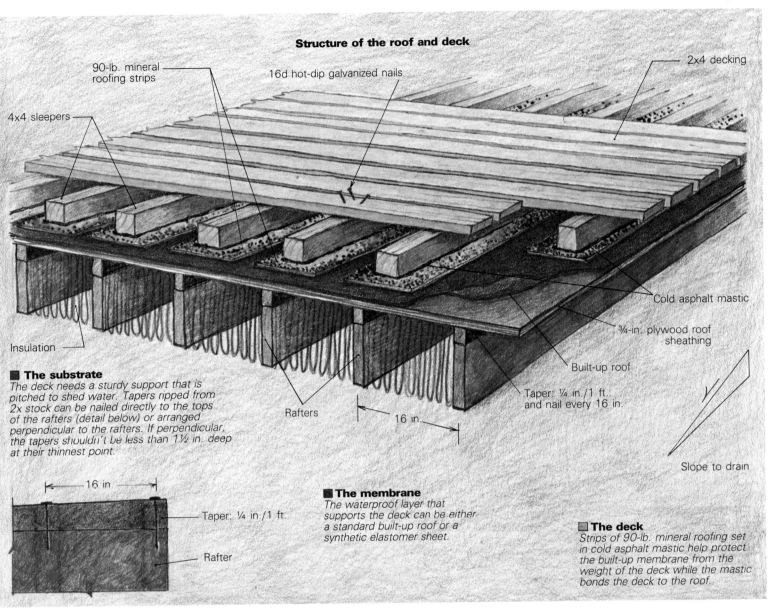

90-lb. mineral roofing strips

16d hot-dip galvanized nails

2x4 decking

4x4 sleepers

Cold asphalt mastic

¾-in. plywood roof sheathing

Built-up roof

Insulation

Rafters

16 in.

Taper: ¼ in./1 ft. and nail every 16 in.

Slope to drain

The substrate
The deck needs a sturdy support that is pitched to shed water. Tapers ripped from 2x stock can be nailed directly to the tops of the rafters (detail below) or arranged perpendicular to the rafters. If perpendicular, the tapers shouldn't be less than 1½ in. deep at their thinnest point.

16 in.

Taper: ¼ in./1 ft.

Rafter

The membrane
The waterproof layer that supports the deck can be either a standard built-up roof or a synthetic elastomer sheet.

The deck
Strips of 90-lb. mineral roofing set in cold asphalt mastic help protect the built-up membrane from the weight of the deck while the mastic bonds the deck to the roof.

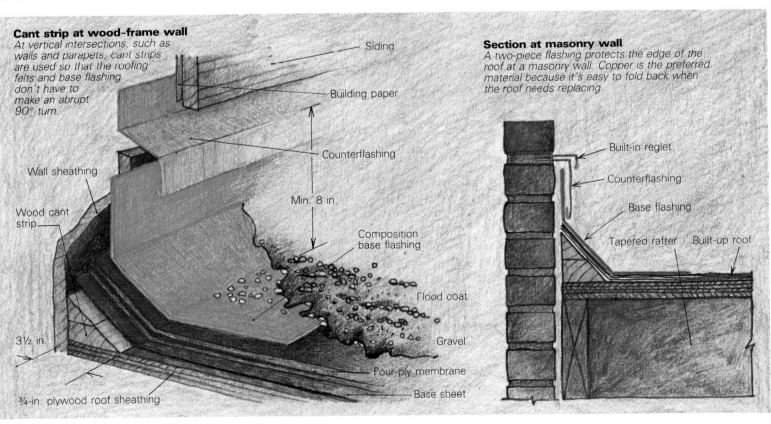

Cant strip at wood-frame wall
At vertical intersections, such as walls and parapets, cant strips are used so that the roofing felts and base flashing don't have to make an abrupt 90° turn.

Siding

Building paper

Wall sheathing

Counterflashing

Wood cant strip

Min. 8 in.

Composition base flashing

Flood coat

Gravel

3½ in.

Four-ply membrane

¾-in. plywood roof sheathing

Base sheet

Section at masonry wall
A two-piece flashing protects the edge of the roof at a masonry wall. Copper is the preferred material because it's easy to fold back when the roof needs replacing.

Built-in reglet

Counterflashing

Base flashing

Tapered rafter

Built-up roof

Illustrations: Frances Ashforth

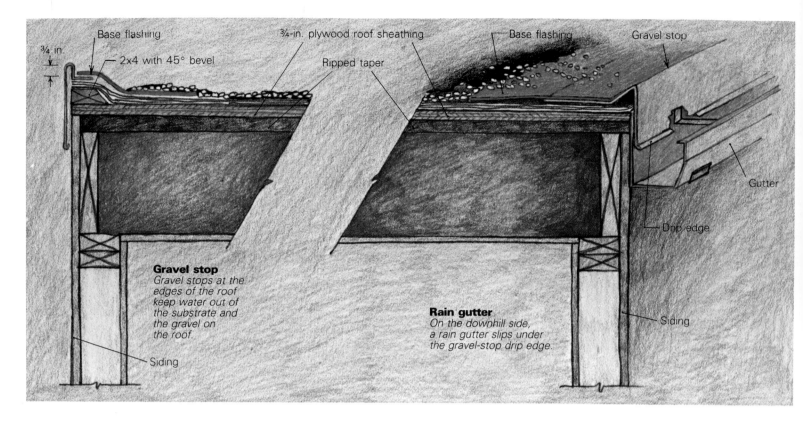

Gravel stop
Gravel stops at the edges of the roof keep water out of the substrate and the gravel on the roof.

Rain gutter
On the downhill side, a rain gutter slips under the gravel-stop drip edge.

Labels: ¾ in. · Base flashing · 2x4 with 45° bevel · ¾-in. plywood roof sheathing · Ripped taper · Base flashing · Gravel stop · Gutter · Drip edge · Siding · Siding

strength, won't rot like the organic ones, and save my worrying about the possible ill effects of asbestos.

Plies are laid successively over the base, starting at the low end of the roof and moving up. The first layer is a 12-in. wide strip butted to the roof edge, which is then covered by a 24-in. wide piece and then a full 36-in. wide sheet. Subsequent full sheets are then shingled over one another to leave 11⅓ in. of exposure.

Roofing felt is bonded with a glazing of hot bitumen between each layer; no two layers of felt should touch anywhere on the roof. A three-ply roof is common here in Kansas, but I always specify four plies. The extra layer costs more of course, but I feel that the added lifespan of the roof more than outweighs the greater cost.

Flashing—According to the National Roofing Contractors Association, the most likely place for a roof deck to develop a leak is at the junction of the horizontal and vertical surfaces. The only way for a builder to prevent these potential leaks is to install the right flashing for the particular condition.

You need two types of flashing on a bitumen membrane roof—base flashing and counterflashing. Base flashing is similar to the nailed-on membrane base, but it's reinforced with fiberglass so that it can bend more easily. It comes in 36-in. wide rolls and has to be cut to the appropriate width. Counterflashing, or cap flashing, overlaps and protects the exposed edges of the base flashing.

When the membrane felts are mopped onto the roof, the roofers will lap the felts over the cant strips and a few inches up the walls. The base flashing covers this intersection and extends at least 8 in. up the wall. In parts of the country where snow is likely to accumulate, 12 in. is better. The extra width will make it

more difficult for the snow to work its way behind the counterflashing.

Once installed, the base flashing is counterflashed with a strip of metal folded into a Z pattern, or with shingles or siding if the exterior wall finish permits. If the wall is masonry, you'll have to let strips of metal flashing into a mortar joint. If you can afford it, use copper. It lasts longer than other flashing materials, and when the roof needs replacing, it's soft enough to be folded back out of the way while the new roof goes on, and then be bent back into place.

At the eaves—The next step is installing gravel stops and rain gutters. Gravel stops come in various profiles. They create a clean, waterproof cap at the eaves and keep the loose gravel on the roof from being blown or washed away. At the downhill edge, a gravel stop should overlap a rain gutter. If your roof has a parapet wall, you'll have to install a metal inset through it for water run-off. Your roofer or sheet metal supplier can make this scupper up in the shop.

Top coat—When the base flashings are down, a thick flood coat is mopped over the entire membrane. Gravel is then added to the molten bitumen. It serves several purposes. The light-colored stones refract and reflect sunlight, blocking out destructive ultraviolet rays. This keeps the roof cooler, and reduces temperature fluctuation. The gravel prevents direct abrasion from the weather, and it acts as a weight to hold the plies on the roof.

The part of the roof under the deck shouldn't have a gravel coating. The deck itself will shield the roof. And you don't want any gravel to get between the deck sleepers and the roof as it will eventually work its way through the membrane and cause trouble. For the same reason, don't walk on the gravel-covered membrane.

Decking—Sleepers should be either pressure-treated wood or the heartwood of a rot-resistant species like redwood. Try to locate them over the rafters and be sure to lay them parallel with the roof slope so that the drainage remains unimpeded.

I used to use 2x4 sleepers laid flat, but I've recently begun to use 4x4s instead, because the extra 2 in. of depth makes it easier to fish out the leaves. The extra wood also lessens your chances of driving a 16d nail through the decking and sleepers into the membrane. Since ¼ in. per ft. is an almost imperceptible slope, I don't taper the sleepers to compensate. But if the slope were any greater, I would taper them to make up for the tilt.

Some roofers may want to hot-mop the sleepers in place, but I would advise against having this done. The sleepers usually rot out before the membrane fails, and hot-mopping them to the roof could mean damaging the membrane when the sleepers have to be removed for replacement.

Instead, I set each sleeper in a bed of cold asphalt mastic on top of a 12-in. wide strip of 90-lb. mineral roofing. This strip, mineral side up, is bonded to the flood coat with the cold mastic. When it comes time to remove the sleepers, the cold joint will give way well before the flood coat. The beauty of this system is that the membrane remains intact while the entire structure is firmly glued to the roof.

For the deck itself, I use either treated southern yellow pine or redwood 2x4s, and I space them ¼ in. apart. If the gap is any larger, things like pencils and envelopes fall through the deck with depressing regularity. I nail each plank to my 4x4 sleepers with three hot-dip galvanized 16d nails. □

Dan Rockhill is assistant professor of architecture at the University of Kansas.

A Decorative Post Cap

Making caps for decks and fence posts

by Robert Vaughan

I make a deck-post cap that has been most popular with my clients. The design isn't particularly original—it's a Williamsburg knock-off—but it does add some curved detail to otherwise boring, rectilinear decks, and it can add a finishing touch to a fence as well. Curves are nice and friendly, unlike the all-too-common detail of simply cutting posts off at an angle.

I'll admit that the caps are time-consuming to make, but once you're set up, a rained-out afternoon can result in 25 or 30 of these profitable details. One building-supply store sold my caps for $7.50 each.

I use pine, pressure-treated with CCA, for the caps. It's widely available around here so I'm able to get some pretty good material. Be aware, however, that the pressure-treating process does not always force chemicals into the heart of a 4x4. That's okay for a post, perhaps, but not for these caps because so much stock is cut away. So select your material carefully. It should have a minimum of checks and should be as dry as possible—the wet stuff cuts poorly and checks easily. There's no doubt in my mind, however, that redwood or cedar would work at least as well, and probably better, than pressure treated pine.

Use straight and square stock. All faces should be the same dimension to ensure consistent results later on. The particular dimensions aren't so important, but make certain that all of the pieces are exactly the same size, particularly in cross section.

You'll need a drill press and a bandsaw. A multi-spur bit works a little better for the boring operations than does a Forstner bit, which has a tendency to burn in the pitchy, stringy pine. A note about safety, however—don't get sloppy while using a drill press or a bandsaw. Awhile back, I cut off my left thumb with a multi-spur bit chucked in a drill press—these days I'm more careful. I use a simple Plexiglas guard to protect the spinning bit when it's at the top of the stroke (and no longer buried in the workpiece). I made it by heating a scrap piece of plastic with a propane torch and bending it over a metal rod.

After cutting the cap using the procedures outlined below, you're ready to install it. Use construction adhesive to glue a 1-in. dia. dowel into the base of the cap. Drill a mating hole in the end of a 4x4 post. Spread a little construction adhesive in the second hole, a little on the bottom of the post cap and slip the cap into place. This will hold the cap firmly; I've found that driving a small finish nail through the base of the cap and into the post risks splitting the cap. □

Robert Vaughan is a professional woodworker in Roanoke, Virginia.

1. Crosscut pieces into 7-in. lengths. The bottom half of the block should be the clearest lumber; that will make the two drilling operations to follow a lot easier. Clamp the block securely to the drill press and bore a 1-in. hole in the bottom of the block. Be sure that all clamps are strong and tight—there's a lot of vibration to overcome when drilling into end grain. Don't hold the workpiece by hand.

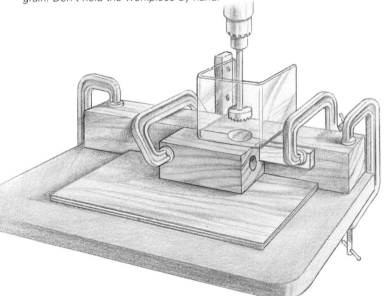

2. Reposition the block and clamp it securely to a thick hardwood fence. The fence should be exactly the same height as the block. Position the drill bit so that the center of a 1⅞-in. hole will fall at the joint between stock and fence. If you have a lot of caps to drill, fasten a stop block to the fence to ensure consistent location of the holes. Drill the first hole clear through the workpiece and into the sub base. Drilling will create a semi-circular trough on the hardwood fence, which will subsequently help to guide the bit. Remove the clamps, turn the block 90° and drill each of the remaining sides.

3. With a marking gauge, tick the centerline of each side of the block near the top. Build a template for the curved lines and mark them on the cap. Note that only one side of the arc needs to be marked. The arc should start at the top of the bored cutout and meet the tick mark about 6½ in. up from the bottom of the block (the ½-in. waste at the top will be useful later as a visual reference). The intersection of the tick mark with the curved lines shows where to stop the bandsaw cut.

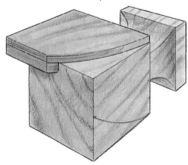

4. Cut each arc, using a stiff blade on the bandsaw (I like a ½-in. blade); use a push stick to keep your hands away from the blade. Stop the cut at the tick mark and back the blade out. Turn the block 90° and cut each of the remaining sides. Separate the pieces. If you want to, sand out any "washboard" from the bandsaw, but don't be too fussy—a finely sanded cap looks out of place on most decks.

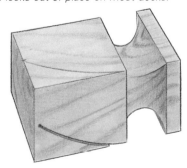

Pattern Routing Exterior Details

This production-line method of shaping helped put a new deck on a Tudor house

by Joseph Wood

"All right, a Tudor!" That's the first thing I thought when I arrived at the Roses' house in Coronado, California. I was there to see about replacing their old deck, and my favorite jobs are outdoor projects where I can re-create custom millwork on period houses.

When I design and build a project for a house with a particular style, whether it's Southwestern, Victorian, Craftsman or whatever, my first step, after looking through my own books, is to go to the library to find out what details define these styles. In this case I was replacing a deck, and my goal was to build one that would look good on a Tudor-style house.

I found that Tudor homes exhibit a heaviness both in construction and in details left exposed. These homes are usually half-timber construction infilled with brick, stucco or stone, and the timbers are big; roofs are steep with wide barge-boards, and colors are often dark.

I did a few things to lend this deck a Tudor quality (photo right). First I sheathed the deck sides to give the profile bulk. For the railing I fabricated broad, pattern-routed balusters of 1x6 redwood. And to create a Tudor theme, I erected an open-roof timber-frame entryway, or pergola. For a heavy look, I stained the deck dark. The finish I used on this deck, and on all my outdoor projects, is Olympic oil-based deck stain. In this case it was Expresso semitransparent stain.

Enclosing the crawlspace—The size and the shape of the deck were restricted by a driveway and a walkway, so I pretty much built the deck with the same perimeter as the old deck. I placed a new continuous concrete footing around the perimeter and built the supporting structure from pressure-treated Douglas fir. The deck joists bear on a 4x6 girder that's supported by a stemwall of pressure-treated 2x4s spaced 16 in. o. c. I sheathed the stemwall, enclosing the crawlspace beneath the deck, to get a heavy look consistent with Tudor architecture. Exposed posts or a lattice skirt would have looked too light. To ventilate beneath the deck, I built mahogany grilles with redwood frames.

All of the exposed stock on this project is kiln-dried clear-heart, vertical-grain redwood, the sweetest wood to work that you could ever find. Redwood is great for outdoor use because it's stable—no warping, checking or splintering—and it lasts a long time. Its workability allowed for a lot of pattern-routed details—most notably the profiled members on the railing and the pergo-

Deck complements house. A timber-frame pergola, closely spaced balusters and an enclosed crawlspace—all made of redwood that's been stained dark—make this deck appear as if it were always part of the Tudor-style house.

la—that really helped the deck work with the Tudor-style house.

Pattern-routed balusters—On the railing I wanted a heavy look in keeping with the Tudor style, but I didn't want a heavy feel. So I used closely spaced 1x6 stock for the balusters and opened them up a bit by cutting each one into a decorative shape. Both bottom and top rails are dadoed to receive the balusters, which are separated with spacer blocks. I attached the top spacer blocks with hot-melt glue. Balusters and bottom spacers float free.

The balusters' heart-shaped design is my own, and I used a router and a template, or pattern, to cut the balusters. Pattern routing is faster and more accurate than cutting with a bandsaw, then sanding the profile edge.

The key to pattern routing is starting with an accurate template. So long as I spend time on fabricating the template, I'm assured of accurate cuts on my workpiece, especially when I'm cutting many identical pieces, such as the balusters.

I make my templates from ¾-in. medium-density fiberboard (MDF) because it doesn't warp, and it's easy to cut and shape. It's also dense, so it holds up to pressure exerted by router-bit bearings. This is important because I often use the templates again on other projects.

My templates start as a sketch of a profile I like on a piece of MDF. I cut the profile with a bandsaw and sand it. To get a perfectly symmetrical template, I cut only one side of the profile. I use this one-half profile to pattern rout one side of another piece of MDF. Then I flip the half-profile template over and rout the other side of the MDF, ending up with the same contours on both sides.

The half-profile template is exactly half the width of the finished template. By marking a line on the finished template along the inside edge of the half-profile template, I can flip it over and position it accurately.

When I'm cutting side details, such as the profile of the balusters, or if I'm removing a lot of material, I first remove the excess with my bandsaw (top photo, right), leaving perhaps ¼ in. of material for my router to remove.

Pattern routing is simple; I attach a template to a workpiece and run a top-bearing flush-trim bit around the template to cut the finished piece (bottom photo, right). Sometimes I tack the template to the blank with a couple of brads; I also make templates that can be clamped to the workpiece without interfering with the router. On this project the baluster templates were tacked in place, but some of the templates I used in shaping the pergola were clamped.

Redwood cuts like cheese, and I can rout practically any kind of pattern I want. But when the router bit comes out of the stock at a right angle to the grain, some tearout or splintering usually occurs. So about ½ in. from the end of the cut, I stop routing, go to the outside of the workpiece and slowly and carefully back the bit into the stock to meet the previous cut.

Building a pergola—The Roses' house, like many Tudor-style homes, features exposed timber framing infilled with stucco—a style called

Getting it close. The baluster profile on the 1x6 redwood was sketched from a pattern, then the excess material was removed to within ¼ in. of the line with a bandsaw.

Getting it perfect. Pattern routing involves running a top-bearing flush-trim bit around a template to cut the workpiece. Here, the baluster pattern is tacked to the workpiece below it.

half-timbering. I've always wanted to try timber framing because it combines both woodworking and construction skills. This deck provided a perfect opportunity because a timber-frame element reinforces the Tudor theme. The pergola is made of heavy redwood stock and features a truss with mortise-and-tenon joinery and a big dovetail at the base of the king post. I did the mortise-and-tenon and dovetail work mainly with flush-trim bits in the router and various templates and jigs. Sharp chisels cleaned everything up.

I used heavy stock—6x6 posts to hold the pergola up, 4x6s for the truss and the carrying beams, and 3x6 rafters—to match the size of the half-timbering on the Roses' house.

I began by setting the posts on anchors in concrete pads and through-bolting the posts to blocking between the joists for stability. Then I fastened the decking and set the pergola's carrying beams in notches at the tops of the posts. At the house end, the carrying beams are mortised into the 1½-in. thick stucco and lag-bolted to the 5/4 sheathing and the first-floor rim joist.

The decorative ends on the carrying beams and on the rafter tails are my interpretation of Tudor: not delicate, but clean cuts, and massive without being square. It took four passes to pattern rout these shapes on the 3½-in. material.

For the first pass I used a ½-in. dia. by ½-in. long top-bearing flush-trim bit, guided by the template (top left photo, above). Then I removed the template and made another pass, this time guiding the bearing on the sides of the previous cut (top right photo, above).

I then changed to a ½-in. by 1-in. top-bearing flush-trim bit and made a third pass, guiding the bearing on the sides of the cut (left photo, facing page). On 4x stock, three passes brought me better than halfway through the workpiece.

Next I cut off the excess stock with a jigsaw, keeping the blade about ¼ in. from the profile line (middle photo, facing page). For the last pass I flipped the piece over and used my 3-hp router with a ½-in. by 2-in. bottom-bearing flush-trim bit, guiding the bearing around on the walls of my last router cut (right photo, facing page). The finished cut is square to the surface and requires no sanding.

With a round-over bit I shaped a decorative profile on the faces of the carrying beams, making two passes, the second one lowered so that the vertical part of the bit cut a shoulder.

Oftentimes, the bearings on routers can heat up and blacken or even seize. So before I use a bearing bit, I give it a shot of bearing lubricant, which I picked up at the local tool outlet.

Fabricating the truss—I wanted an uncluttered truss to match the spacing of the half-timbering on the house, so I designed a simple king-post truss with two webs (drawing facing page). I drew the truss full size on my workbench to get the angles and the dimensions.

To cut the profile of the 4x4 king post, I used the same pattern-routing technique I used to cut the balusters.

I used templates I made for an earlier project to shape the dovetail that connects the king post to the bottom chord. When pattern routing hard an-

Routing thick stock

Bit rides against template. With a rafter-tail template clamped to the 4x6 rafter stock, the author guides a ½-in. flush-trim, top-bearing bit along the template's edge to begin shaping the end profile.

Bit rides against previous pass. Using the same router setup but with the template removed, the author guides the bearing along the cut made by the first pass; in effect, the rafter becomes its own template.

gles, such as the narrow portion of the dovetail and the wide portion of the dovetail mortise, the bit leaves a curved portion equal to its radius, and this curve must be squared up with a chisel.

A bridle joint connects the king post and the top chords; that is, the top chords are forked to fit over tenons in the king post. The lower ends of the top chords have long mortises that accept the tenoned ends of the bottom chord.

Although the heavy top-chord timbers aren't likely to sag over their short 7-ft. 6-in. spans, I added 2x4 webs for stability and looks. I cut the mortises and the tenons for the webs traditionally, with a handsaw and a chisel.

Then I put the truss together. I used pegs and construction adhesive to secure all of the tenons and construction adhesive only in the dovetail. I tapped all of the pegs in a little and put in button plugs for a nice finished appearance.

I set the truss on top of the posts and the carrying beams. Four long dowels—one in the top of each 6x6 post and one into each carrying beam—hold the truss in place, and the whole thing was further secured with a countersunk lag screw into each carrying beam.

The pergola's ridge sits in a half-lap at the top of the truss and butts into the stucco. Once the ridge was in place, I set the three pairs of 3x6 rafters. These aren't as tall as the 4x6 truss chords, so the 10, 2x3 purlins are let into the truss chords ¾ in. The final task was installing the pendant (see sidebar) under the trusses' king post.

Imagining my forefathers' approval for my first timber-framing project was one of the most satisfying moments of my building career. □

———————

Joseph Wood, a fourth-generation builder, owns Wood's Shop, a design/build company in San Diego, Calif., that specializes in creative outdoor projects and traditional building restoration. Photos by Rich Ziegner except where noted.

Making finials

One late night I happened to catch an old British film on television. In this film there was a scene in a Tudor-era church, and while the plot was unfolding, I was eyeing the architecture. Sure enough, I saw at the top of a newel post a knoblike ornament, called a finial, that I could use on the deck's railing. I used the same design for a pendant directly below the truss' king post.

I cut both the finials and the pendant from 6x6 redwood stock on a bandsaw and narrowed the base to 4 in. by 4 in. on a table saw. The only difference between the finials and the pendant is that I trimmed the base of each finial with redwood brick molding.

First I made a pattern of the finial profile from ¼-in. plywood. Then I traced the profile on a piece of redwood and cut it on the bandsaw. To profile the rest of the sides, I couldn't trace the pattern on the final, so I tacked to the workpiece a length of ¼-in. plywood with the outline of the finial drawn on it (photo below). Then I cut the rest of the finial. —*J. W.*

Drawings: Vince Babak

Longer bit, deeper pass. A 1-in. top-bearing bit also rides against the routed portion of the rafter tail to bring the cut about halfway through the 3½-in. thick redwood stock. Harder woods will require shallower passes.

Flip the piece over and cut off the excess after tracing the decorative end profile on the face of the rafter using the original template. The author rough-cuts to within ¼ in. of the outline with a jigsaw.

Bit rides against routed profile. The author cleans up the rough cut using a more-powerful (3 hp) router with a 2-in. flush-trim bottom-bearing bit, again using the stock to guide the bit. The finished end requires no sanding.

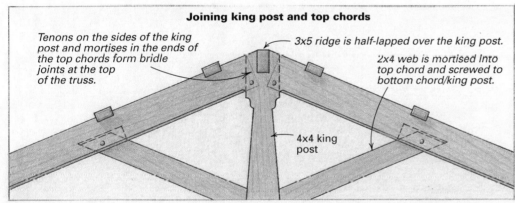

Joining king post and top chords

Tenons on the sides of the king post and mortises in the ends of the top chords form bridle joints at the top of the truss.

3x5 ridge is half-lapped over the king post.

2x4 web is mortised Into top chord and screwed to bottom chord/king post.

4x4 king post

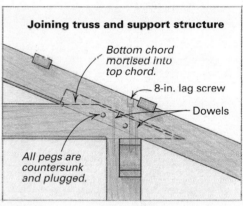

Joining truss and support structure

Bottom chord mortised into top chord.

8-in. lag screw

Dowels

All pegs are countersunk and plugged.

Assembling a king-post truss

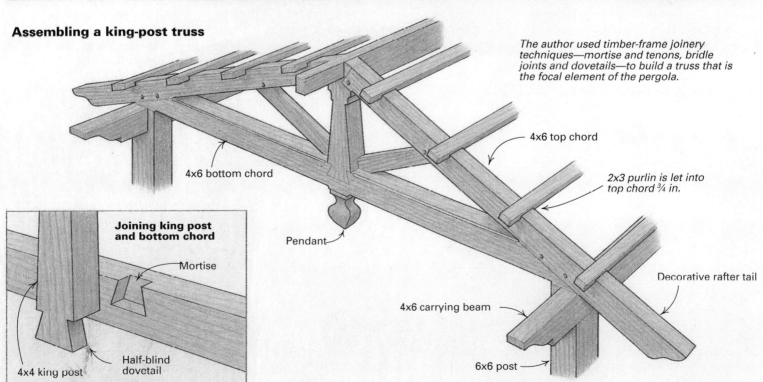

The author used timber-frame joinery techniques—mortise and tenons, bridle joints and dovetails—to build a truss that is the focal element of the pergola.

4x6 top chord

2x3 purlin is let into top chord ¾ in.

4x6 bottom chord

Pendant

Decorative rafter tail

Joining king post and bottom chord

Mortise

4x4 king post

Half-blind dovetail

4x6 carrying beam

6x6 post

Railing Against the Elements

With careful detailing and proper flashing, exterior woodwork should last for decades

by Scott McBride

Mushrooming problems. **When fungi sprout from railings, it's a rotten sign. Pictured here is a close-up of the railing the author was hired to replace.**

The moist climate of New York's Hudson Valley isn't exactly ideal for carpenters like myself. We have muggy summers, slushy winters and three months of rain in between. As we struggle through weeks of unremitting precipitation with fogged-up levels and wet chalk lines, there is but one consolation: come April, a billion fungal spores will bloom, reaching into every water-logged mudsill, fascia and doorstep. That means a guaranteed crop of rot-repair jobs in the coming season.

Of all the woodwork exposed to the elements, none is so vulnerable as the white pine porch railing. With the right combination of faulty detailing and wind-driven rain, a railing can be reduced to shredded wheat in about eight years. I typically rebuild several of these railings each year. In this article I'll describe one such project (photo facing page).

Getting organized—A railing around the flat roof of a garage had rotted out (photo above), and the owner asked me to build a new one. This wasn't a deck that was used, so the railing was decorative rather than functional, and one of my worries was installing the new railing without making the roof leak (more about

that later). On a house across town, my customer had spotted a railing that he liked and asked if I could reproduce it. I said I could and took down the address.

Before leaving the job site, I made a list of the rail sections I would be replacing, their lengths and the number of posts I would need. Later that afternoon I found the house with the railing my customer liked, strolled up the walk and began jotting down measurements. The family dog objected strenuously to my presence, but no one called the police.

Back at the shop I drew a full-scale section of the railing and a partial elevation showing the repeating elements. The next bit of work was to make layout sticks (or rods) showing the baluster spacing for the different rail sections (see the sidebar on p. 96). This would tell me the exact number of partial- and full-length balusters that I would need.

The right wood—I'm fortunate to have a good supplier who specializes in boat lumber. He carries premium grades of redwood, cedar, cypress and Honduras mahogany, all of which resist decay well. I used cypress for the rails and balusters because it's less expensive and

because the rough stock is a little thicker than the others. Cypress is mostly flatsawn from small trees, though, and the grain tends to lift if the wood isn't painted immediately.

Much of the Western red cedar at this yard is vertical grain—the annual rings run perpendicular to the face—and hence inherently stable. I typically use it for wider pieces where cupping could be a problem. Square caps on posts fit this description because they are so short in length. They should always be made from vertical-grain material or they'll curl in the sun like potato chips.

For this job I bought 2-in. thick cypress for the rails and balusters, 5/4 red cedar for the rail caps and post caps, and 1-in. cedar boards for the box newel posts.

Bevels and birds' mouths—After cutting the lumber into rough lengths with a circular saw, I jointed, ripped and thickness-planed the pieces to finish dimensions. Using a table saw, I beveled the top, middle and bottom rail caps at 15° (drawing facing page). Besides looking nice, the bevels keep water from sitting on what would otherwise be level surfaces.

I had to cut a bird's mouth on the bottom of

Railing anatomy

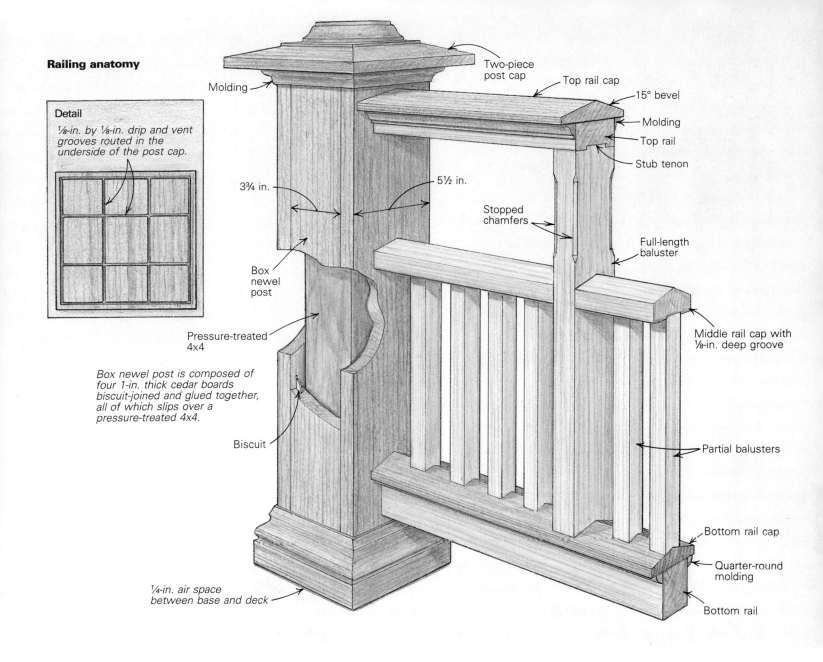

Molding

Two-piece post cap

Top rail cap

15° bevel

Molding

Top rail

Stub tenon

Detail
⅛-in. by ⅛-in. drip and vent grooves routed in the underside of the post cap.

3¾ in.

5½ in.

Stopped chamfers

Full-length baluster

Box newel post

Pressure-treated 4x4

Middle rail cap with ⅛-in. deep groove

Box newel post is composed of four 1-in. thick cedar boards biscuit-joined and glued together, all of which slips over a pressure-treated 4x4.

Biscuit

Partial balusters

Bottom rail cap

Quarter-round molding

¼-in. air space between base and deck

Bottom rail

each partial baluster so that the baluster would fit over the beveled cap on the bottom rail. Rather than make each bird's mouth individually, I first crosscut my 2x6 baluster stock to finished length and jointed one edge. I then ripped just enough off the opposite edge to make it parallel to the jointed edge. With the blade on my radial-arm saw raised off the table and tilted 15°, I made an angled crosscut halfway through the thickness of the baluster stock. Flipping the piece, I made the same cut from the opposite face. Individual balusters were then ripped from the 2x6, with each bird's mouth already formed.

I used the same setup to cut birds' mouths in the full-length balusters, but cut them one at a time because they were wider. These uprights also have decorative stopped-chamfers routed into them. The chamfer bounces light smack into your eye in a most appealing way.

Next I set up the dado cutterhead on my table saw. I ploughed a shallow groove for the partial balusters in the underside of the middle rail, and another to receive the full-length balusters in the underside of the top rail. Although these grooves are a mere ⅛ in. deep, they made assembling the railing much easier

Decorative railings. The original railing around the deck over the garage rotted out prematurely, so the owner of the house commissioned a new railing of a slightly different design—one copied from a house across town. Carefully detailed of cypress and cedar, the new railing should last a long time. *Photo by Kevin Ireton.*

and ensured positive alignment of the vertical members. The tops of the partial balusters are housed completely in the groove, but the tops of the full-length balusters also have stub-tenons, cut on the radial-arm saw.

Box newel posts—Each box newel post is a two-part affair. I cut rough posts from pressure-treated yellow pine 4x4s and outfitted them with soldered-copper base flashing (top drawing, next page). This flashing would later be heat-welded to the new roof surface (more on that in a minute). A finished cedar box newel post would slip over the rough post and receive the railings.

Because the sides of the box newels were to be butt-joined, two sides were left their full 5½-in. width, and the other two were ripped down to 3¾ in. so that the finished post would be square in cross section. I saved the rippings to make the quarter-round molding that's under the bottom rail cap.

The box newels were glued up with a generous helping of resorcinol to keep out moisture. Glue also does a better job of keeping the corner joints from opening up than nails do. To align the sides of the box newels dur-

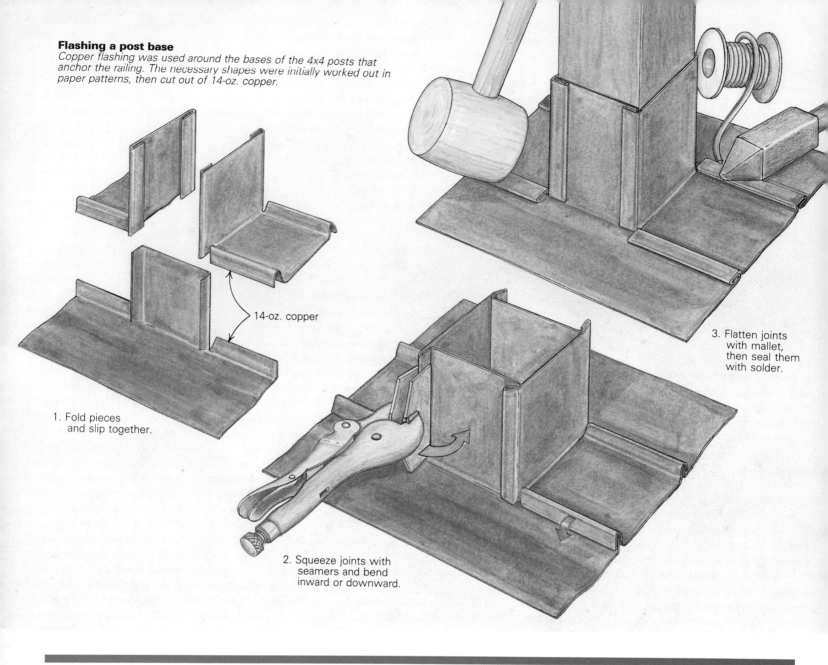

Flashing a post base
Copper flashing was used around the bases of the 4x4 posts that anchor the railing. The necessary shapes were initially worked out in paper patterns, then cut out of 14-oz. copper.

14-oz. copper

1. Fold pieces and slip together.

2. Squeeze joints with seamers and bend inward or downward.

3. Flatten joints with mallet, then seal them with solder.

Spacing balusters

Spacing balusters correctly is a simple trick, but it's surprising how many carpenters are stumped by it. The object is to have equal spaces between each baluster and between balusters and posts.

Suppose your railing is $71\frac{5}{8}$ in. long., your balusters are $1\frac{1}{2}$ in. square, and you want a spacing of 4 in. o. c.—that is, $1\frac{1}{2}$ in. of wood, $2\frac{1}{2}$ in. of air, $1\frac{1}{2}$ in. of wood, and so on.

Begin the layout from the post on the left, by tentatively laying off a $2\frac{1}{2}$-in. space ending at point A in the drawing below. From there you stretch a tape and see how close you come to the post on the right with a multiple of 4 in. In this case 68 in. (a multiple of 4) brings us to point B, which is $1\frac{1}{8}$ in. from the right-hand post. Forget about this remainder for the moment.

Dividing 68 by 4 tells you that you will have 17 intervals containing one baluster and one space, plus the extra space you laid out at the start. That makes 17 balusters and 18 spaces. Now you're going to lay out the combined dimensions of all 17 balusters ($17 \times 1\frac{1}{2}$ in. = $25\frac{1}{2}$ in.) from the left-hand post. That brings you to point C. From there you're goint to lay off the combined dimensions of all the spaces, bringing you to point B—close, but not close enough. To land directly at the post, divide the remaining $1\frac{1}{8}$-in. by 18 (the number of spaces) and add the quotient to each space. How fortunate $1\frac{1}{8} \div 18 = \frac{1}{16}$. That gives you a nice, neat adjusted dimension for the space of $2\frac{9}{16}$ in.

If the numbers don't divide evenly, I'll use a pair of spring dividers to find the exact space dimension by trial and error, stepping off the distance from Point C to the post. When I find the right setting, I lay off the first space. Then I add this dimension to the baluster width, reset the spring dividers to the sum distance, and step off the actual spacing. This method avoids the accumulated error that happens when using a ruler and pencil, not to mention all that excruciating arithmetic. —*S. M.*

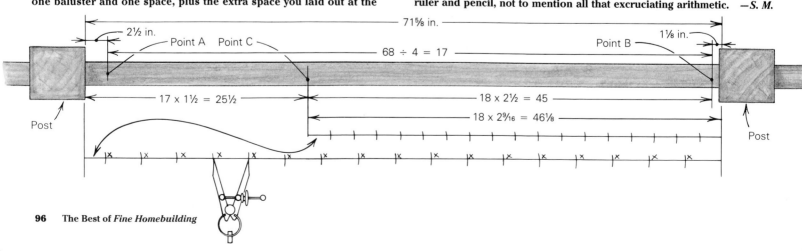

$71\frac{5}{8}$ in.

$2\frac{1}{2}$ in.

Point A   Point C

$68 \div 4 = 17$

$1\frac{1}{8}$ in.

Point B

Post

$17 \times 1\frac{1}{2} = 25\frac{1}{2}$

$18 \times 2\frac{1}{2} = 45$

$18 \times 2\frac{9}{16} = 46\frac{1}{8}$

Post

ing glue-up, I used three biscuit joints along the length of each side.

The last parts to be fabricated were the post caps. The square lower part of the cap is a shallow truncated pyramid, produced by making four consecutive bevel cuts on a table saw. I made the round upper part on a shaper. The two parts were glued together with the grain of each parallel to the other, so they would expand and contract in unison.

To reduce the amount of water running down the face of the post, I cut a drip groove around the underside of the cap with a ⅛-in. veiner bit mounted in a router table. Using the same bit, I also routed a series of ventilation grooves into the underside of the cap in a tic-tac-toe pattern (detail drawing, p. 95). They allow air taken in at the bottom between the rough post and the box newels to escape at the top without letting in rain.

Assembling the rail sections—I assembled the rail sections in the shop where I could count on dry weather and warm temperatures. The first step was to face-nail the bottom rail cap to the bottom rail. I placed the nails so that they would be covered subsequently by the ends of the balusters.

I toenailed the balusters in place with 4d galvanized finish nails. This was easy to do because the bevel and bird's-mouth joinery prevented the balusters from skidding around as I drove the nails.

The middle rail was cut into segments to fit between the full-length balusters. I took the lengths of the segments, along with the spacing for the partial balusters, directly from the layout on the bottom rail cap. Every other segment of the middle rail could be attached with 3-in. galvanized screws through the uprights. The intervening segments were toenailed with 8d galvanized finish nails. Then I face-nailed through the middle rail down onto the tops of the partial balusters. The top rail was screwed down onto the full-length balusters. I left the top-rail cap loose so that it could be trimmed on site for a tight fit between the newels.

Still in the shop, I caulked all the components with a paintable silicone caulk, primed the wood and then painted it with a good-quality latex house paint. With a truckload of completed rail sections and posts I headed for the job site with my crew.

Installation—My roofing contractor had replaced the existing 90-lb. rolled roof over the garage with a single-ply modified-bitumen roof. One advantage of modified bitumen is that repairs and alterations can be heat-welded into the membrane long after the initial installation. This meant I didn't have to coordinate my schedule with that of the roofer to fuse the copper base flashings to the new roof. Flashing strips of the bitumen were melted on top of the copper flange (photo right), providing two layers of protection (including the copper) around the base of the post and a "through-flashed" layer of roofing beneath the post.

The railing is U-shaped in plan, and I an-

Fused flashings. One of the advantages of a single-ply modified-bitumen roof is that repairs and alterations can be heat-welded to the membrane long after the initial installation. Here, flashing strips of bitumen, which cover the copper base flashings around the 4x4 posts, are being fused to the roof membrane.

chored the two ends into the house. However, I didn't want to penetrate the roof membrane with framing or fasteners, so the newel posts are only attached to the deck by way of their flashings. Although this method of attaching the posts provides superb weather protection, the intermediate posts are a bit wobbly. This was okay for this particular deck because the railing is strictly decorative. Where a roof deck is subject to heavy use, the posts should be securely anchored to the framing.

With the rough posts in place, installation of the railing was straightforward. Box newels were slipped over the 4x4s and roughly plumbed. We stretched a line between corner newels and shimmed the intermediate newels up to the line. All box newels were held up off the deck at least ¼ in. to allow ventilation. The difference between the rough post dimension (3½ in. by 3½ in.) and the internal dimension of the box newels (3¾ in. by 3¾ in.) allowed for some adjustment. When the newels were just where we wanted them, we simply nailed through the shims and into the rough posts.

In some cases rail sections had to be trimmed. When the fit was good, top, middle and bottom rails were snugged up to the posts with galvanized screws. All that remained was to glue cypress plugs into the counterbored holes and shave them flush. □

Scott McBride is a builder in Sperryville, Va., and a contributing editor of Fine Homebuilding. *Photos by author except where noted.*

A Comfortable Outdoor Bench

A canted back and a sloped seat make this built-in bench beautiful, weather resistant and a restful place to sit

by David Bright

A good carpenter always takes pride in a well-done job, but some projects are especially gratifying. Such was the reward that Mark Schouten, Doug TeVelde and I got from building a red-cedar bench on a deck overlooking the placid waters of Lake Whatcom near Bellingham, Washington (photos facing page). The bench was designed by David Hall.

The bench had to meet three criteria: It had to be truly comfortable; it had to survive in a damp, rainy climate; it had to complement its spectacular setting and yet blend with it.

Many wood benches are uncomfortable. They make you slouch forward, or they make you wish you could. Some benches cut you behind the knees or across the back. But this bench is made for comfort. The back leans out 10° from vertical, and the seat dips in a curve about ¾ in. below horizontal—just enough to allow your back to rest naturally against the rails.

As well as making the bench comfortable, the slopes and the curves diminish the number of horizontal planes on the bench and allow few places for rainwater to collect. To further the bench's weather resistance, the wood was spaced to allow air to circulate, and all fasteners were galvanized or made of stainless steel.

Back and seat supports—Before assembling the bench, all the pieces of clear red cedar used in the construction were run through a surface sander. The extra effort spent dressing the wood in the beginning saved considerable time at the end when it was time to apply the finish to the bench.

The bench frame consists of seat supports and back supports. The back supports are made of 2x4s. Before attaching them to the deck, we dadoed the supports to accept the 1x4 back rails. The back supports are canted 10° from plumb and, as the drawing above shows, we cut the bottom ends of the back supports to fit against the

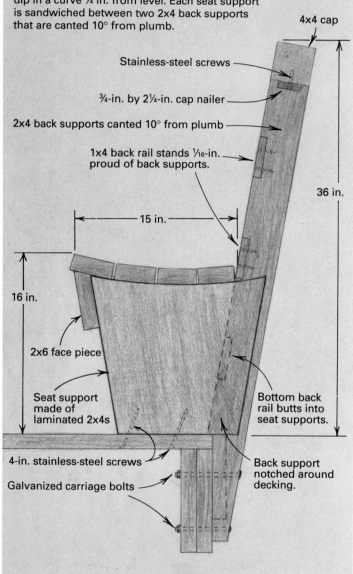

Built for comfort. The seat supports dip in a curve ¾ in. from level. Each seat support is sandwiched between two 2x4 back supports that are canted 10° from plumb.

- 4x4 cap
- Stainless-steel screws
- ¾-in. by 2¼-in. cap nailer
- 2x4 back supports canted 10° from plumb
- 1x4 back rail stands ⅟₁₆-in. proud of back supports.
- 36 in.
- 15 in.
- 16 in.
- 2x6 face piece
- Seat support made of laminated 2x4s
- 4-in. stainless-steel screws
- Galvanized carriage bolts
- Bottom back rail butts into seat supports.
- Back support notched around decking.

deck frame, where two galvanized carriage bolts and a little construction adhesive hold them solid and tight. Two back supports sandwich each seat support.

The seat supports are made of 2x4s laminated with exterior urethane resin glue. To attach the seat supports to the deck, we angled 4-in. stainless-steel screws through the decking from underneath. Throughout the assembly process we tried to fasten either from underneath or on a

vertical surface; our idea was to avoid areas where water could penetrate the wood through the holes made by the screws. We also wanted to avoid showing any fasteners.

We also ran a horizontal 2x6 face piece around the inside perimeter of the seat. The face piece is flush with the top edge of the seat support and was beveled (on a table saw) to continue the curving top plane of the seat support. The face piece helps finish the design of the bench; it also supports the first row of 2x4s that make the seat.

Curves, cants and compound miters—The first row of 2x4s that make the bench's seat extends over the face piece about ½ in. The 2x4s were installed beginning at the inside edge—the edge that your knees rest against. Working toward the back rest gave us room to fasten screws into the miter joint at the bench's two inside corners. If we had started from the outside and worked inward toward the front, the previous row of mitered 2x4s would not have allowed us to fasten the next row. A good adhesive caulking seals the end grain of the miter and helps keep the joint tight.

If the seat had no curve, mitering its joints would have been a matter of cutting good, familiar 45s. But the curve demanded a compound miter, and a different one for each row—a fact that will no doubt be lost on the folks who will sit there. On the inside row the miter must be undercut: on the second, not so much; the third is almost perpendicular; and the last miter is shorter on the surface than on the bottom. The safest way to get the right compound cut was to use two scrap pieces of 2x4 to test the angles. You may draw like Frank Lloyd Wright and calculate like Einstein, but when it comes to making that critical cut, you'd better test first.

If you're off a bit, you can change the angle of your blade or remove some wood with a chisel or a block plane. Remember, red cedar is soft, and it's easy to overcorrect. (You know the

Drawing: Chuck Lockhart

Comfort that stands up to the weather. By slanting or curving most of the surfaces on this bench, the builders were able to build a seat that is truly comfortable. The slopes and the curves also eliminate horizontal planes on which standing water can collect.

story: Soon your board is too short, and your temper is shorter.)

The horizontal back rails are 1x4s, which were carefully let into the canted 2x4 back supports. The rails extend about 1/16-in. past the back supports. Because of the 10° angle of the back, the corner miters are compound angles, too. Construction adhesive and two nails from each direction hold the corners tight. To avoid splitting, we predrilled the corners before nailing. The mitered corners seem to hang in midair, but they are as strong as can be.

Before the rail cap was installed, a 3/4-in. by 2 1/4-in. cedar strip running the length of the back rest was notched flush into the top of each 2x4 back support. This strip provided a way to screw from underneath into the 4x4 cedar cap. The cor-

ners of the cap were also mitered at a compound angle very similar to the first miter of the seat.

After the cap was installed, we touched up the wood with 150-grit sandpaper. For comfort's sake, we made sure all the sharp edges of the bench were eased with sandpaper. Sanding also removes nicks and pencil lines that are unavoidable during construction. We didn't want these small imperfections to get sealed into the wood. After sanding, the bench was sealed with a semi-transparent stain.

It's not often that people in our business get to build a project so perfectly suited to its location. We did our best, and the results rewarded us. □

David Bright is a carpenter and writer in Lynden, Wash. Photos by Charles Miller.

Building a Sauna

Shop-built components ensure precise work and ease of assembly

by Ric Hanisch

With one match I light the wick of the kerosene lantern, then the paper and kindling in the potbellied stove. In 20 minutes the air temperature in this tiny room will rise from 40°F to 110°F, and I'll return to throw on some more chunks of whatever is left over from recent building projects and shop scrap. Another wait, and the temperature climbs to a toasty 180°F and the wooden surfaces of walls, floors, ceiling and benches are coming into equilibrium with the air. Time to bake. Sauna time!

People the world over use heat alternating with cold for cleansing and restorative baths. The Finnish version, called sauna, was particu-

Ric Hanisch lives in Quakertown, Pa. Photos by the author.

larly attractive to me and my family. In use for over 1,000 years, the Finnish sauna requires a hot, relatively dry atmosphere to induce sweating—lots of it. This is followed by a cooling-off period—a plunge into a lake, a roll in the snow or a cool shower. Traditionally this sequence is carried out two or three times, and the intensity of the contrast is certainly invigorating. The ritual of firing up the sauna and its potential as a lively, healthful, contemplative family event drew us to this form of bath.

Having seen a number of saunas, I wanted to design one that we could build ourselves. A convenient source of fuel made a wood-fired sauna practical for us, and this in turn pointed to a separate structure that wouldn't take anything else with it if it went up in flames. The basic ele-

ments would be simple—a woodstove for heat, a small, weathertight shell with ample benches, and a cold-water line for cooling, since we have no lake on our property. We began to look for an appropriate site as we were developing our vision of the project. A group of boulders where the mowed lawn funnels into a woodland path embraced a tiny site—just right for our sauna. A neighbor's goat, staked out for a few days, cleared the brambles and verdant growth of poison ivy that covered the area.

Time, materials and money—We wanted to spend no more than two months on the sauna. There seems to be an optimal amount of time during which an owner-built project can continue to sustain enthusiasm. We had about $500

Prefabrication in the shop. **To minimize damage to a beautiful site and to achieve more precise joinery, Hanisch decided to prefabricate the floor, walls and roof sections in his workshop. At left, the frames for a gable wall and a sidewall are test-fitted on top of the floor. Above, a custom door and its trim are installed, along with beveled cedar siding. When all the components are nailed down, the siding will be trimmed flush.**

worth of materials on hand, and what we had to buy came to about $1,300.

Even though the site was within reach of an extension cord, I made an early decision to prefabricate the sauna's floor, wall and roof sections in my barn workshop. There were several reasons why this made good sense. First of all, I liked the idea of placing the sauna on its site. Stick-by-stick construction at the site would have reduced the natural carpet of mayapples and jack-in-the-pulpits to a trampled mess. I wanted the site to recover quickly.

The possibility of prefabrication was also intriguing because it presented a different set of design parameters. For example, shop assembly allowed trim details throughout the sauna to be more precise than they would have been had we built on site. And a prefabricated design would enable me to construct saunas in kit form, deliver the parts to a client's site and assemble each structure with a minimum of on-site labor.

Many people had a hand in the construction. I did most of the framing and precision work with Mike Reed, a skillful and energetic hired helper. A lot of the shingling, siding and insulation was done by family and friends who agreed to forego wages in exchange for lifetime sauna privileges.

Fabrication sequence—We framed the floor first, using 3x4 fir joists set on 16-in. centers. At each corner, we bolted on 6x6 preservative-treated yellow pine posts 12 in. long. These would later bear on a stone foundation. Spruce T&G 2x6 decking went over the floor framing.

We used the completed floor as a work platform to frame up the slanted sidewalls (photo above left). I had seen outward-slanting sidewalls on Finnish log saunas and also on a sauna built in California by a friend. The slanting walls reduce the volume to be heated while maintaining sufficient bench space, and they also provide comfortable back support.

The gable walls and the sidewalls were framed with 2x4s. I designed the sidewalls so that their top plates would overlap the gable-wall framing. This keying of parts created a strong joint and also made assembly easier (bottom left photo, next page).

Using some redwood salvaged from an old water main and some catalpa that I had in the shop, I made a frame-and-panel door for the sauna, using dark redwood for rails and stiles and catalpa for panels (photo, facing page).

Once all the walls had been framed and test-fitted on the floor, we set to finishing them off (photo above right). Door and window frames were installed, along with insulation and finish walls. Against the inside edges of the studs, we stapled silver Mylar. This thin sheet material cuts down on air leakage and also acts as a radiant barrier, helping to keep heat in. For interior walls, I used 1x6 T&G Philippine mahogany. Surprisingly, this knot-free, moisture-tolerant stock was less expensive than any clear domestic T&G material I could find. On all interior walls, we left the bottom-most piece of paneling off so that we would be able to screw the bottom plate to the floor. All site connections were made with 3-in. drywall screws.

Against the outside edges of the studs, we nailed beveled white cedar siding. On the sidewalls, the siding was cut back to leave room for a single corner board, which was installed after the sauna was assembled.

The two matching roof sections have three rafters apiece, two of which bear directly on the gable-wall framing. Each rafter consists of a redwood filler strip sandwiched between a pair of fir 3x4s (bottom left photo, next page). Yellow pine 2x6 T&G decking was installed over the rafters. Yellow pine often contains knots and

pitch pockets that might bleed under high temperatures, but so far this hasn't been a problem.

Silver Mylar was stapled down over the roof deck, and this was followed by a layer of tar paper and horizontal 1x2 strapping. The strapping creates an air space beneath the cedar roof shingles, allowing them to dry out after a rain. When shingling each roof section, we left off the course closest to the ridge so that the 2x decking could be screwed together where the roof sections meet. Where the rafters meet at the ridge, a screwed redwood spline took the place of the filler strip, tying opposite rafters together.

Putting it together—After installing the floor, we mustered an enthusiastic bunch of volunteers to put the remaining parts together. Before transporting the gable walls, I protected their windows by tacking flakeboard panels over the trim. By 10 o'clock both gables were in place (top photo, next page), and we had the shell done by lunch. We later shingled the ridge of the roof, completed a few remaining trim details, and installed the stove. The stainless-steel flue passes out of the sauna through a transite-glazed window in the rear gable wall.

The deck provides a private space to relax and cool off between heats. Its shape also anchors the sauna to the site, provides screening and determines an approach to the sauna that threads between boulders.

The treated 2x8 frame of the deck rests on a stone foundation. For the decking, I used the same 3x4 fir that had gone into floor framing and rafters, but resawed it on my bandsaw into 1½-in. by 4-in. planks. I've found that a bandsawn surface is less likely to splinter than a planed surface, and its texture provides much firmer footing. The deck's treated 4x4 posts hold horizontal supports for a screen of recycled plaster lath. □

Assembly on the site. The structure is supported by preservative-treated 6x6 posts, bolted to the corners of the floor and resting atop rock pillars at the site. On raising day, above, the gable-end walls go up first. The glass in the windows is protected with flakeboard panels that have been temporarily nailed to the trim. The bottom board has been left off the interior of the gable wall, revealing a silver Mylar radiant barrier. This makes it possible to screw the plate to the floor deck. Two-by-four lap joints at the top corners of the walls were designed as strong, easy field connections, as shown below left. Twin roof sections have three rafters apiece, with each rafter built up from two fir 3x4s that sandwich a redwood filler strip. The two outer rafters bear directly on the gable framing. A single length of corner-board trim runs along the edges of each sidewall, as shown below right. Shingling the roof ridge and installing the stainless-steel flue complete the job.

A Traditional Finnish Sauna

Keeping in the heat with well-chinked logs and a sod roof

by Stanley Joseph

Because of our weather, seacoast and lakes, we in Maine feel akin to the Finns—especially to their ancient method of steam-bathing, the sauna, taken in a bathhouse or room with the same name. The earliest Finnish sauna was a large hole dug into a hillside. Rocks were mounded inside around a stone post-and-lintel structure with a fire beneath it. A smokehole dug in the hill cleared the smoke from inside. After the rocks were heated, the sauna entrance was closed off and the smokehole was plugged. Then the bathers entered and doused the rocks with plenty of water. The bathers would sweat and scrub themselves, jump in a nearby lake or roll in the snow, then run back to the steamy cave. Bathing could easily go on this way for an hour or more.

Evolving from this pit was the log sauna with a smokehole and an open firepit. The *savu-sauna*, or smokesauna, got its name because the inside of the sauna was smoke-blackened and smelled of smoke. A few purists in Finland still prefer the smokesauna, but more popular is the log sauna, with a masonry or iron stove and chimney and a sod roof. This is the type of sauna that I set out to build with my friend Steve Hanson, a woodworker and tree surgeon who lives nearby.

Logging time—Logs are the perfect material for a sauna because they conduct heat slowly. And we had a ready source of logs on our property. We had no question about where to build—on a tiny island in the center of a half-acre pond. So in the spring of 1985, we girdled the chosen trees by peeling away bark and layers of cambium from the circumference of each tree. This prevents the sap from rising and kills the tree, seasoning it as it stands. In the spring of the next year, Hanson felled the trees, which I then peeled and hauled to our work site next to the pond.

We began to build the sauna in the fall. In a 12-ft. by 12-ft. square, we dug down about one foot, sloping the ground toward one corner. I laid a 6-in. drainpipe from this corner to the pond to carry away the water from washing and cleaning inside the sauna. We spread polyethylene over the 12-ft. square, then laid down a foot-deep layer of crushed rock. Large, flat stones laid at the corners of a 10-ft. square would act as a foundation. The finished sauna would measure 10 ft. square

Before the sauna begins, the author cuts a hole in the ice and fires up the Finnish stove that fits into a cast-concrete frame in the log wall.

outside and 8½ ft. square inside, with 8 ft. from floor to ceiling height at the ridge.

Joining the logs—We decided to join the logs with half-dovetailed notches because of the tight, strong corner joint they make and for their traditional appearance (top photo, next page). This joint looks complex when assembled, but each log end requires only two angled notches. Viewed from the end of the log, the top notch slopes downward from the interior to the exterior while the bottom notch slopes upward. Both notches have the same pitch (see *FHB* #13, pp. 57-59 for details of how to cut a similar half-dovetail notch).

With a chainsaw, Hanson cut the ends down to a 6-in. thickness and rough-cut the dovetails. Then I used a drawshave to smooth the notches and cut the decorative wedge at each log end. After Hanson finished off the dovetails, we hand-carried each log across the narrow bridge to the building site.

We set and leveled each log and secured it with an 8-in. galvanized spike at each end. We left a gap of about an inch between each course—enough for a good amount of chinking. For the first round of logs, we used cedar because of its resistance to decay; the remaining logs were spruce. The logs in each round became progressively smaller in diameter as they progressed from foundation to ridge, diminishing from 12 in. to 8 in.

Hanson cut 3-in. tenons on the ends of each log around the door opening. The tenons were let into a continuous mortise in the 6x8 posts that frame the door, then pegged in place (top photo, next page). The bottom log that forms the sill was joined to the 6x8 posts by full blind mortise and tenons.

A through-the-wall stove—After laying five rounds of logs, we finished off the opening that we had left for the stove and its concrete frame. The frame supports the stove and protects the logs from the heat (top photo, below). The stove, which was custom-made by an old stovemaker from Finland, was designed to be set in the outside wall of the building so that it can draw fresh air directly. Additionally, we can feed the fire from the outside.

We cast the concrete frame flat on a piece of plywood, then tilted it into place. The frame is held in place by 2x4 splines that key into a continuous mortise cast into the outside edges of the frame and a matching mortise cut into the surrounding logs. We fit the stove into the concrete frame, sealing between them with a ¼-in. thick fiberglass gasket and a coat of refractory cement that I mixed with water and applied with a trowel. This fiberglass gasket is often used on lobster boats to protect wood decking from the heat of the engine's muffler. Only the exterior door of the stove is visible from outside. Welded to the stove inside are baskets for holding rocks, which in our sauna are granite and soapstone.

We chinked the logs in the following spring using the method described by Charles McRaven (*FHB* #26, pp. 48-51). We attached metal lath between the logs on the outside, stuffed fiberglass batting against the lath and nailed on another layer of lath on

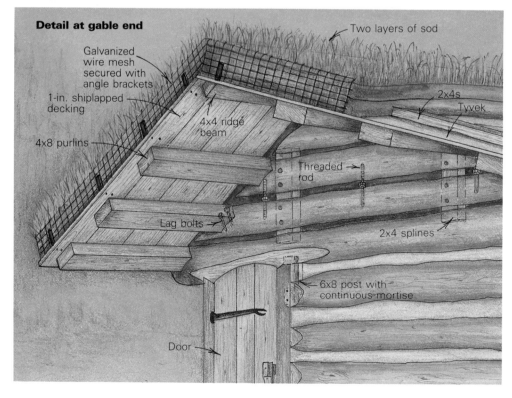

Detail at gable end

The logs of the sauna are joined with half-dovetails (above), traditional in their look. Steve Hanson checks the fit of a window frame, which is mortised into the logs at the sill and held in place at the jambs by a spline let into the logs. A temporary wedge holds the logs apart before the spline is pegged to the logs. The concrete frame will support a woodstove, protecting the logs from heat.

Oversized stringers (above) will carry a three-tiered bench. Before bench slats are installed, stringers are rounded to a comfortable curve.

the inside. A 1:2:3 formula of cement, lime and sand gave the mix a pleasing light-grey color and more plasticity than would a formula with a higher percentage of cement.

Eaves of grass—For the roof structure we used 4x8 spruce purlins and a 4x4 ridge beam cut on a neighbor's sawmill. Hanson let the purlins into the gable walls, spacing them 2 ft. o. c., and secured them to the log walls with lag bolts (drawing facing page). To further stabilize the gables, we mortised short 2x4 splines through the gable-end logs. We also tied the log that acts as a door lintel to the log above with ⅝-in. diameter, 8-in. long threaded galvanized rods. Washers and nuts helped to maintain the 1-in. space between the two logs. The shiplapped 1-in. thick spruce roof sheathing was planed on one side and dried thoroughly. The smooth side of the sheathing is exposed on the inside, so we painted on a protective mixture of boiled linseed oil and turpentine, mixed 1:1. On the outside, the generous overhangs all around protect the logs from the weather.

Sod roofs are used on traditional Finnish saunas. On these roofs, lapped layers of birchbark were used between sheathing and sod. As a substitute, we covered the sheathing with three layers of Tyvek. The Tyvek prevents water from passing into the sheathing but allows water vapor to pass freely between the sauna and the roof. We built an 8-in. galvanized-wire sod-stop and a drip edge along the eaves and supported it with angle brackets (drawing facing page). Where the brackets are bolted to the roof, we sealed the Tyvek and sheathing with Bituthene Ice & Water Shield, a self-adhering bitumen membrane (manufactured by W. R. Grace Construction Products Division, 62 Whittemore Ave., Cambridge, Mass. 02140).

Dupont doesn't recommend Tyvek for use under wood-shingle or sod roofs simply because it's slippery and is therefore dangerous to work on. To get around this problem, we built a framework of 2x4s that we hung over the peak and covered with sod. To help keep the sod from sliding off of the roof, we had set the roof pitch at 14°. This also encouraged water to soak into the sod, and helped snow to stay on the roof for extra insulation.

The best sod is said to come from a field that already has a good growth of grasses, clover and vetch because each of these has a dense root system. So in the spring we burned off an area of a grassy field to encourage the growth of new grasses and followed up with a 10-10-10 fertilizer, lime and a sowing of mixed grasses, vetch and clover. Over the summer I mowed the area regularly. In late August, we cut the sod into 1-ft. squares. We pulled up the sod with a tractor and loaded it into a pickup truck for transportation to the edge of the pond, where we gently transferred it to a wheelbarrow for the short trip across the bridge to the sauna.

The sod squares were the right size for easy handling. We used two layers of sod, each 4 in. thick. The first layer was laid grass side down; the second, grass side up. The upside-down layer discourages roots from the top layer from growing down into the roof, and as the lower layer decomposes it gives the top layer organic matter to live on. The top layer of grass began to grow after the first rain. After two years the sod is alive and well—even after extremely dry summers (photo below). We've never watered the sod ourselves and have found that its 8-in. thickness enables it to survive drought. I've thought about adding eave and rake boards to the roof, though, because the wire mesh allows moisture to wick from the edges of the sod too quickly. I wish I had followed the traditional Finnish method of using partially hollowed logs fit to the edges of a sod roof as combination rake boards and gutters.

Finishing up—The sauna door swings open to the outside. This is a safety feature all saunas should use to provide easy exit in an emergency. We arched the top of the 30-in. by 72-in. door (photo below). Doug Wilson, a blacksmith on Deer Isle, Maine, made a set of strap hinges. Several coats of spar varnish finish the door.

Inside, we built three tiers of benches, each 16½ in. apart. The width of the lowest bench is 18 in., the middle 20 in., and the top 24 in. All three benches run the length of one wall and the top bench turns to run along an adjacent wall. The bench structure

looks like a wide stair with deep risers and treads (bottom photo, facing page). To make the large stringers for the stepped bench, we angled the ends of 2x6s, hung them on the wall with joist hangers and rested the lower ends on large, flat rocks. We built up the stringers with 1xs and 2xs, rounding their outside corners to make a comfortable shape for sitting. To make the seats, we nailed on 1x3 spruce slats with a 3/16-in. gap between slats, set the nails, and coated the slats with the 1:1 oil and turpentine mixture. We finished up with a floor of slats, spaced and nailed to a 2x4 frame that rests on the gravel bed. To protect the floor, we stacked up 4-in. thick rough-cut granite pavers around the stove.

We celebrate the elements—Building our sauna took a lot of time, but that's what it's all about—slowing down and celebrating the elements: earth, fire and water. Imagine a grey day in January, 10° F outside. The sauna was fired up three or four hours ago and a hole cut in the ice of the pond. At dusk, when the inside of the sauna is 180° F, the bather enters. Ahhh. The body relaxes, the pores open, sweat drops in profusion. After 10 or 15 minutes the bather rushes out in a cloud of steam, plunges into the black hole in the ice and emerges with a primal yell. Then it's back into the sauna to repeat the ritual. There's nothing like it. □

Stanley Joseph is a farmer and writer in Harborside, Maine. Photos by Lynn Karlin.

The sod is held in place by a galvanized wire-mesh stop that runs around the perimeter of the roof. From spring to fall, a ladder remains against the sauna to allow the author access to the roof for handmowing with a scythe and fertilizing with seaweed, and for an occasional sunbath.

Japanese-Style Bath House

Western materials and methods combine with Eastern design in this backyard retreat

by Kip Mesirow

Every summer I spend some time at Tassajara Hot Springs in the Santa Cruz mountains of California. I stay at a Zen Buddhist monastery there, which is open to guests for part of the year. I always take along my carpentry tools on these visits, and I fix this and that in exchange for my keep.

Tassajara is a small place, and over the years I have often met up with Hank and Jill Wesselman. We'd see each other on the trails or in the baths that are fed by natural hot springs. Hank and Jill would sit back in the baths, pleasantly simmering and appreciating the Japanese influence on the surroundings—the trees, rocks and water courses there are all organized into subtle relationships that balance and enhance one another. Hank fancied having his own Japanese bath, and having seen my work, asked me if I'd like to build one. The house he and Jill had just bought in the Berkeley hills had the perfect backyard for a little Japanese-style outbuilding. The wooded lots surrounding theirs create a private alcove that seems far from neighbors.

The Japanese bath—While Westerners have always preferred the solitary soak, the Japanese bathe in groups, enjoying the economy of one large wooden tub full of hot water and the camaraderie of shared relaxation time. When Westerners first became aware of Japanese customs toward the end of the 19th cen-

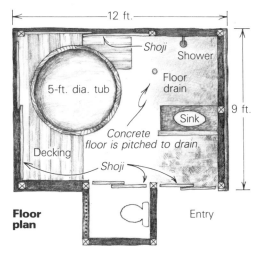

Floor plan

12 ft.

Shoji — Shower

5-ft. dia. tub

Floor drain

Sink

Concrete floor is pitched to drain.

Decking

Shoji

9 ft.

Entry

The bath house basks in the afternoon sun in this view from the Wesselmans' back door. Above the entry, a bamboo gutter directs rainfall to the side. The copper roof jack is a plumbing vent, and has its own gable roof.

Photo: Hank Wesselman

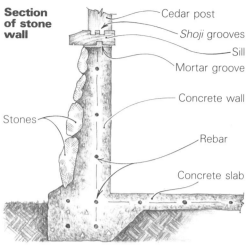

Section of stone wall

Cedar post
Shoji grooves
Sill
Mortar groove
Concrete wall
Stones
Rebar
Concrete slab

tury, many were appalled by the lack of hygiene implied by a communal bath. Wouldn't a bunch of sweaty bathers pollute a common body of water? But as Edwin Morse pointed out in his 1885 portrayal of Japanese life, *Japanese Homes and Their Surroundings,* nothing could be farther from the truth. Most Japanese, especially the working classes, would bathe two or three times a day, and they would first wash themselves with a separate bucket of water and a towel. The communal bath was just for soaking.

The Wesselmans wanted to build an updated version of a classic Japanese bath house. Instead of the separate bucket, it would have a sink and a nearby shower for tub prep. They also wanted a toilet in the building. Given these requirements, the floor plan began to take shape.

In keeping with traditional Japanese buildings, I chose a hipped gable with a shallow pitch (3-in-12) for the roof form. Vents at the gables would allow for air circulation. We all wanted the bath house to emphasize simplicity and subdued materials, so the building is mostly wood and stucco. Since the inside of the roof would be exposed to view, I wanted contrasting woods to punctuate the differences between rafters and roof decking, so I chose redwood decking and Port Orford cedar rafters, beams and blocking. I drew it all up with 2-ft. eaves hovering over a 7-ft. high wall, and showed the plan to the Wesselmans. They liked it, and the project was on.

Timber preparation—As soon as I had the lumber quantities calculated, I ordered a load of roughsawn Port Orford cedar from a mill in Oregon. The timbers needed time to adjust to the Bay Area climate, so as soon as the lumber arrived I stacked and stickered the 2x6s and covered them with a plastic tarp. I took the 6x6s aside, and ripped a slot to the center of each one. For the posts I cut this slot into the least attractive side. For the beams, I cut the slot into the crown side. Then I drove hardwood wedges, 3 ft. to 4 ft. o. c., into these slots. Each week my partner, Gary Wilson, and I would drive these wedges a little farther into the slots or replace them with thicker ones. This Japanese technique lets the wood dry evenly from the inside as well as the outside,

Cast-in-place stonework covers the 3-ft. concrete walls that surround the bath house on three sides (top). Cedar sill plates are bolted to the walls and are half-lapped at the corner on the left. The post above the lap is mortised into the sill while the one on the right is secured to the concrete by horizontal foundation bolts cast into the wall. Black bamboo grillwork covers the vent in the lavatory wall. Above, a helper holds a rock in place while Mesirow drills holes in the form boards for the tie wire. The slab floor has been finished with a pebble texture to avoid a slippery surface in a wet location. The recessed floor at the entry will be covered with slate.

Joinery details

The sills that bolt directly to the slab on the west wall of the bath house were cut from 3x10 white cedar stock, and vary in width according to the interior wall finish. They are beveled for drainage on the exterior side, and hang over the edge of the slab by about 1½ in. Each post slips into a 2x2 mortise, cut 2 in. deep into the sills.

A stepped tenon at the top of each post fits into a corresponding mortise in the top plates. For a post this size, the steps are equal to the width of my Japanese framing square—9/16 in. The width of the tenon shoulders is equal to 30% of the post width. The tenon itself is equal to 40% of the post width. The mortise is flared slightly at the top to allow wedges driven into the post to expand the tenon, locking the post in place. The top plate is just a little wider than the post, so the post can meet the plate at its chamfered edges.

The intersection of the top plates at a corner is the most adventurous joint in the building, and we made a test model (photo at right) before cutting into the long stuff. The junction is a lap joint, but it's complicated by a diagonal groove that accepts the hip rafter. Instead of simple square sides, the laps are chamfered and recessed into the neighboring plate. This trick keeps light from shining through the joint if the wood shrinks. —K. M.

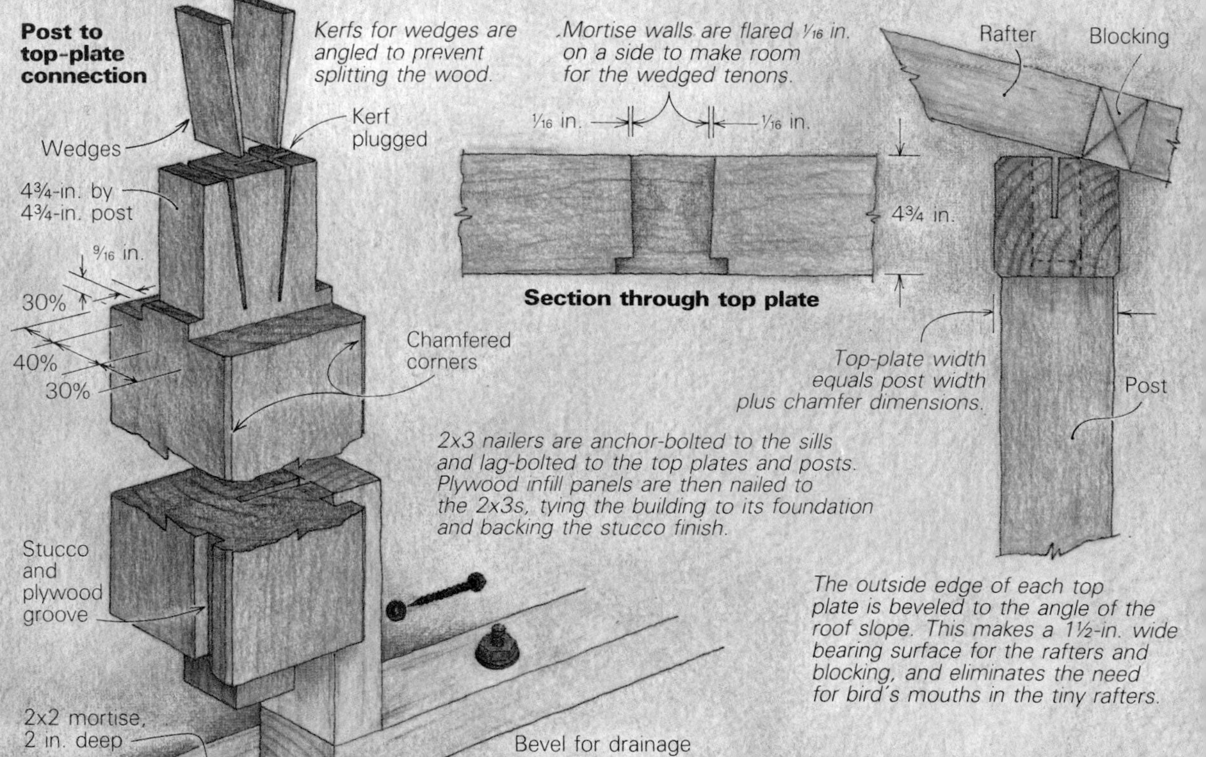

Post to top-plate connection

Wedges

4¾-in. by 4¾-in. post

9/16 in.

30%
40%
30%

Chamfered corners

Stucco and plywood groove

2x2 mortise, 2 in. deep

Concrete

Drip kerf

Post-to-sill connection

Kerfs for wedges are angled to prevent splitting the wood.

Kerf plugged

Mortise walls are flared 1/16 in. on a side to make room for the wedged tenons.

1/16 in. ⊣⊢ ⊣⊢ 1/16 in.

Section through top plate

2x3 nailers are anchor-bolted to the sills and lag-bolted to the top plates and posts. Plywood infill panels are then nailed to the 2x3s, tying the building to its foundation and backing the stucco finish.

Bevel for drainage

Rafter   Blocking

4¾ in.

Top-plate width equals post width plus chamfer dimensions.

Post

The outside edge of each top plate is beveled to the angle of the roof slope. This makes a 1½-in. wide bearing surface for the rafters and blocking, and eliminates the need for bird's mouths in the tiny rafters.

Mesirow built this actual-size model of the top-plate intersection before cutting the full-length versions. The tapered diagonal notch is for the hip rafter. The wedged kerfs in the top of each piece promote even drying and help prevent checking.

and reduces the possibility that the visible faces will check. As they dried, the 6x6s deformed slightly so they were no longer square. But they were large enough that once they were dry we could true them up to 4¾ in. square. One post is visible from all four sides, so we picked out the best heart-center one that we could find and didn't slot it. We just kept it in a shady spot and hoped it wouldn't twist and check—luckily, it didn't.

A stone-faced foundation—We wanted a band of rocks around the base of this little building to broaden the range of textures between the various materials and to create a natural transition between the site and the walls. A stone foundation would have been just fine, but stone foundations have to be over-engineered in this earthquake zone, and so they are expensive. We compromised: concrete walls with stone veneer.

We poured a standard 6-in. slab floor and reinforced it with rebar rather than wire mesh. This meant that it could better withstand the load of the hot tub. At the perimeter we left rebar stubs every 18 in. so we could tie the slab into the walls.

Next we laid out our stones on the ground in front of the future walls. We used a rock called coldwater canyon stone by the supplier. We chose it for its texture and color, and

because the stones are thin enough to leave plenty of room for the concrete behind. Once we had a stone pattern that fit together with minimal gaps, we staked the outer form-boards in place. To give the building a sturdy look, we sloped these forms inward to give a batter of 5 in. in 3 ft.

Once the outer forms were completed, we used 16-ga. tie wire to secure each stone to the inside of the forms—best side facing out (photo previous page, bottom). We'd never done this before, and we had the nagging feeling that wet concrete would want to flow through the gaps between the stones and obliterate their weathered faces. To guard against this, we stuffed strips of foam rubber into each one of these cracks. The local foam-rubber mattress shop was happy to give us their scraps for this task.

With the stones wired in place, we put up the inner forms and filled them slowly with concrete. We used a slightly wet mix of about 4½-in. to 5-in. slump, delivered to the form in buckets. Each bucket load was carefully tamped into place with a rod to make sure that all the voids were filled.

We let the concrete cure in the forms for three days, and watered them twice a day to make sure the concrete didn't dry out too fast. On the fourth day, we pulled away the forms and the foam rubber. We wire-brushed away

the concrete that had gotten through the foam. The technique had worked. The foam-rubber dams did their job, and the stones were unblemished and thoroughly bonded to the wall (photo previous page, top).

In the shop—When the foundation was complete, we turned to our pile of cedar. We pulled the wedges out of the 6x6s and trued them up to 4¾ in. square on our thickness planer. Then we labeled them with their locations in the structure and cut grooves for plywood and stucco. Posts and beams adjacent to the door and window openings were grooved to receive *shoji* screens. We also cut two grooves for *shoji* in the 3x10 cedar sill plates that rest atop the concrete walls next to the tub. We leveled these sills on a bed of mortar spread across the top of the concrete walls. A third groove in the bottom of the sills acts as a keyway for the mortar. The two sills are lapped at the northern corner and mortised for the intersecting post.

Framework—In keeping with traditional Japanese buildings, this bath house has a post-and-beam superstructure. The sills and top plates are mortised to receive the posts, which have tenons on both ends. More joinery details are shown above.

Once we had the wall members cut out and

Illustrations: Christopher Clapp

Detail of roof eave

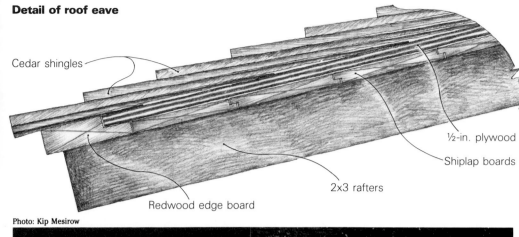

Cedar shingles

Redwood edge board

2x3 rafters

½-in. plywood

Shiplap boards

Photo: Kip Mesirow

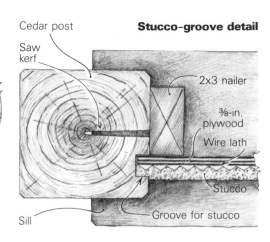

Cedar post

Saw kerf

Stucco-groove detail

2x3 nailer

⅜-in. plywood

Wire lath

Stucco

Sill

Groove for stucco

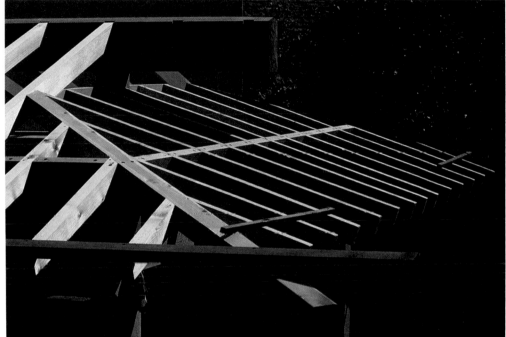

Countersunk drywall screws tie rafters and blocking to the top plates and the ridge beam. The four missing rafters at the gable end will be installed after the shingles are extended to the gable vents.

the eaves, and the plywood strengthens the roof and prevents the roofing nails from coming through the redwood planks.

Walls—The infill walls are tied to 2x3 nailers that align with the grooves in the posts (drawing, above right). On the outside, a layer of ⅜-in. plywood overlaps the nailers and fits into the grooves in the posts. This provides shear strength and backing for the stucco finish. Traditional Japanese post-and-beam buildings have a matrix of wooden lath holding together earthen walls, but we felt taking that route would be too time-consuming. Although we were striving for a finely wrought, traditional structure, we had to keep the building on a realistic course.

Stucco was our choice for the walls, and we used it both inside and out. We covered the plywood with kraft paper and expanded-metal lath, and troweled on a total of three coats. The last coat had extra-fine sand in it to make it easier to get a smooth finish, and we mixed in some beige pigment to reduce the glare common to the pure white mix. After each coat we softened the edges between the stucco and the wood with a wet paintbrush, and we wiped away any stucco that slopped onto the posts or plates. We also misted the stucco two or three times a day for three days, and covered the fresh walls with plywood to protect them from the direct sun. We are far from stucco experts, but these measures seem to have helped—no appreciable cracks have developed since the walls were finished.

The tub—We had the tub built in place because there weren't any openings large enough to bring it into the room—the tub is 5 ft. in diameter and 3½ ft. high. Although I've participated in the assembly of several tubs, I hired an expert to put this one together, and I'm glad I did. I hung around and helped him, which lowered the assembly price, and also gave me a chance to learn more about the cooper's art. Redwood tubs aren't easy to assemble if you don't know what you're doing, and they almost invariably develop a leak or two. We haven't had trouble with this one yet, but I've got a phone number to call if we do.

The plumbing was straightforward, and we tied the hot-water stubs for both the sink and the shower to a Paloma demand gas water heater. The hot tub has its own gas-fired unit.

marked, we ripped the 2x6s into 2x3s for rafters, and then used the thickness planer to bring them down to 1¾ in. by 2¾ in. At this point all the posts, beams, plates and rafters were finished by hand with Japanese planes, including the chamfers on each edge. When we were done, we wrapped each stick with butcher paper to protect it from fingerprints and discoloration from the sun.

It took two of us three weeks to prepare the assorted framing members. Once we got all the pieces to the site, it took only four hours to put them together.

Building the roof—In traditional Japanese construction, the ridge beam is supported by a series of king posts, which rest on beams that span the walls from one top plate across to the other—something like ceiling joists in ordinary stick framing. These beams are sturdy and run perpendicular to the ridge beam. As a variation on this design, we ran a single beam under the ridge beam and parallel to it (photo next page). This lower beam joins the top plate at each end with a tapered dovetail joint.

Once we had this beam in position, we knocked in the king posts and lowered the

5x9 ridge beam onto them. The ridge beam has notches 1 ft. o. c. to cradle the ends of the rafters (photo above). We countersunk two 3-in. drywall screws into the ends of each rafter at the ridge, and another one where the rafter meets the top plate. We used screws so the structure wouldn't have to withstand the pounding that comes along with nails, and so that if anything went wrong, it would be easy to take things apart without tearing them up.

We put up all the common rafters first, then the 3x3 hip rafters, followed by the hip jack rafters. The complexity and small size of the roof gave us some problems where the two vents went into the gables. Because there was so little room to work, we had to put down our decking and shingles before we could continue with the last pairs of rafters that complete each gable.

At the eave line, a redwood edge board supports the first row of shingles. These pieces are 3½ in. wide and 1 in. thick. The decking on the underside of the roof is redwood, ½ in. thick and rabbeted to a shiplap pattern. We covered this redwood skin with a layer of ½-in. plywood to bring us flush with the edging, as shown in the drawing above left. This detail gives the building a generous reveal at

A collar tie with three king posts positions the ridge beam and distributes loads to the end walls. Above it, the redwood decking and cedar framing members offer a rich contrast to the stuccoed walls and *shoji* screens. Mesirow also made the copper light fixture and candle holder.

The slate entry—The first step into the bath is onto a small field of square slate tiles. These natural pavers are durable and pretty, but slate isn't an easy material to cut. Since we had to groove the slates for the *shoji* doors, we were on the lookout for such a method. We eyed our tile cutter's water saw, and found our answer.

We put each slate on the tile saw's movable bed and made several ¼-in. deep passes. We moved the slate over a little after each pass. This made a rough groove that we cleaned up later with a small carborundum stone. When we set the slates, we used a straightedge to align the grooves.

The *shoji* doors have Port Orford cedar frames with red cedar panels at the bottom. Instead of rice paper, both the doors and the windows have fiberglass panels. Rice paper doesn't last long when exposed to rough weather and cats. The window *shoji* slide on waxed oak runners pressed into grooves in the 3x8 sills. The door *shoji* were to run in the grooves in the slate, but every little bit of rock or sand that got into the track would cause the doors to make a terrible grating sound as they slid along. Some of the larger bits of rock even scratched grooves into the slate. To solve this problem, we mortised a pair of steel wheels into the bottom rails of each door. Now the doors roll along the track with little resistance. Still, the window *shoji* on the oak tracks have the smoothest action, and they slide almost soundlessly.

Air flow—Without adequate air circulation, a bath house will grow mildew and mold, and the wooden framing will rot. We put vents in the gable ends to ward off these evils, but they didn't work—surfaces just stayed wet when the *shojis* were closed. But if a *shoji* was open for a while, things dried out. We'd neglected to provide an opening for air to enter when the building is closed. Fortunately there is a vent in the toilet alcove that can supply the needed outside air when the lavatory door is left partially open. It's an imperfect solution, but better than a screen door somewhere else. Next time I'll plan for such a vent.

Now that it's finished—It took six months and $25,000 to build this bath house, and through it all the clients remained cheerful and understanding. We all discussed problems as they came up, and worked out solutions that made everybody happy. All through the construction process, Gary Wilson and I assumed that our ability to innovate and adapt would get us through the unforeseen situations. But I've learned that this is not the Japanese way of building. Every piece in a Japanese structure is designed to fall within a module of some sort, therefore the size of one element, such as a post, affects everything else. For instance, in Western building it's not considered a problem to end up with a trimmed row of tiles in the corner of a room. In Japan, however, a trimmed row of tiles would be considered a mistake. In Japan, responsibility for keeping track of these myriad dimensions rests firmly on the shoulders of the layout man, and it's no coincidence that it takes about 20 years to get good at it. These wizards travel from site to site, and their job is to mark where everything goes.

So we don't have any delusions about the pedigree of this building. But bathing at night in the soft glow from the copper light fixtures, with *shoji* open to the moonlight on the shrubbery and the woodwork reflected on the water, everything seems quite alright. □

Kip Mesirow is a builder and partner in HIDA Tool and Hardware Co., in San Rafael, Calif.

Pondside Gazebo

Monolithic frost walls and cantilevered beams support a structure on wet, unstable soil

by Alan Guazzoni

When I was asked to design a gazebo near a pond in Stowe, Vermont, the first thing that came to mind was the trite image of a white Victorian structure with ladies under parasols strolling to music from *The Music Man*. There was no way, however, of reconciling this image with the 1960's contemporary home that the gazebo was to serve. Instead, the gazebo's classical, yet abstracted form (photo above) was heavily influenced by the roof pitches, the general massing and the clapboard siding of the main house.

One of the first problems in this project was siting the building. The client wanted the gazebo and its surrounding deck close to the pond to

serve as a cabana and swimming platform. But she did not want to obscure the view of the pond from the house. So we built a full-scale plywood profile of the gazebo and tried it in several locations before agreeing on the best one. At this point we also decided to lower the grade of the building site 4 ft. to minimize the building's height and to create a more intimate relationship between the deck and the pond.

The sunfish-shaped deck that would surround the gazebo was to be supported on a parallel pair of 15-ft. long concrete frost walls, with one portion of the deck cantilevering 10 ft. and extending over the pond. Dropping the elevation of the

building site placed the top of the frost walls only 12 in. above the pond. Furthermore, one frost wall would not only be 4 ft. from the water, but that wall's footing also would be 4 ft. below the pond's surface.

The soil conditions were not ideal for building, and the mechanics of placing the concrete walls and footings created a potential for water flooding the excavation. But the high level of moisture in the soil and Vermont's cold climate made frost considerations paramount, so the frost walls and wide footings seemed a logical approach. After talking it over with the contractors, Dave Buchanan and Jeff Holden of Holden

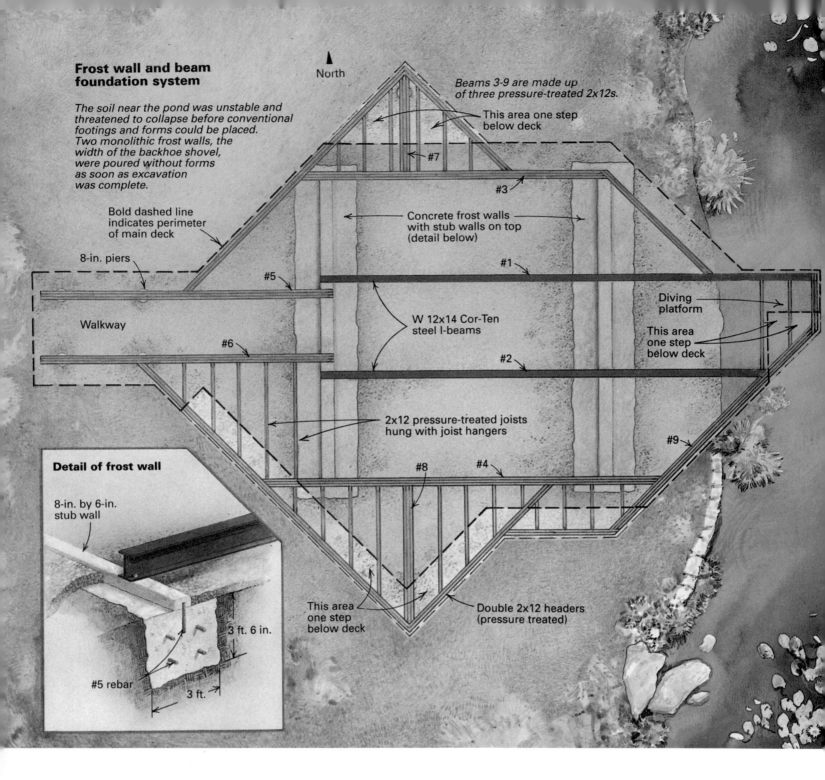

Frost wall and beam foundation system

The soil near the pond was unstable and threatened to collapse before conventional footings and forms could be placed. Two monolithic frost walls, the width of the backhoe shovel, were poured without forms as soon as excavation was complete.

North

Beams 3-9 are made up of three pressure-treated 2x12s.

This area one step below deck

#7

#3

Concrete frost walls with stub walls on top (detail below)

#1

Bold dashed line indicates perimeter of main deck

8-in. piers

#5

W 12x14 Cor-Ten steel I-beams

Diving platform

This area one step below deck

Walkway

#6

#2

2x12 pressure-treated joists hung with joist hangers

#9

Detail of frost wall

8-in. by 6-in. stub wall

#8

#4

#5 rebar

3 ft. 6 in.

3 ft.

This area one step below deck

Double 2x12 headers (pressure treated)

Construction Company, we decided to take our chances with placing the concrete and try to solve any related problems as they cropped up —draining the pond would be the last resort.

Digging in unstable ground—Lowering the grade of the building site proved uneventful. The soil appeared dry and stable. It was only while digging down an additional 4 ft. for the frost-wall footings that the inevitable became apparent. Although water did not flood the trenches, the excavation walls were saturated and became extremely unstable. The trench walls collapsed nearly as fast as the backhoe could dig. The situation was further complicated by a huge boulder uncovered about 3 ft. down in one of the trenches. The size and the location of this boulder made it impossible to remove.

Dave Walker, a local concrete wizard, realized that the conventional approach of forming and pouring footings and foundation walls would not work here. The area was becoming a vast mud pit. Walker's solution was simple and elegant. He and his crew poured two parallel, monolithic frost walls, without forms, by filling the trenches with concrete as soon as they were dug. With a concrete truck standing by, each footing trench was dug 3 ft. wide—the width of the backhoe shovel—and 4 ft. deep where possible. The trenches were then filled a third of the way with 3,000 psi concrete, and two horizontal #5 rebars were placed the full length of the trenches. More concrete was poured, filling the trench about two-thirds of the way, and two more #5 horizontal rebars were installed.

Walker and his crew then poured the remainder of the concrete, bringing the top of the wall up to grade. They placed #5 rebar vertically 2 ft. o. c. along the length of the frost walls to serve as dowels for stub walls. This work was done in minutes, before the trench walls could collapse. Once poured, the concrete held back the earth, which in turn held in the concrete.

The next day, Walker's crew poured 8-in. wide by 6-in. high continuous stub walls the length of each frost wall over the #5 rebar dowels (detail, drawing above). They placed ¾-in. steel anchor bolts in the stub walls so that they could bolt down the steel I-beams. At the same time, the crew poured eight 8-in. piers, using Sono tubes, to catch the extremities of the deck, the steps and the walkway. With the concrete work completed, Walker and his crew shaped and compacted the area to drain surface water away from the frost walls.

Cantilevered beams—The main deck and the walkway are supported by six longitudinal beams running on an east-west axis (drawing above). The steps on each side of the deck are

Drawing: Gary Williamson

supported by two lateral beams running on a north-south axis. The outboard ends of these beams are supported by 8-in. piers. A single angled beam and variously angled headers enclose the deck, the steps and the diving platform.

All of the beams are made up of three pressure-treated 2x12s, except beams #1 and #2, which are W 12x14 steel I-beams. All beam headers are doubled-up, pressure-treated 2x12s. The primary support for the deck and its 10-ft. cantilever is provided by beams #1 and #2, which are bolted to the two frost walls. The gazebo itself acts as a counterweight to the cantilever, so special beam seats were not necessary.

Because the beams are exposed to the weather continuously, we used Cor-Ten Type A steel (from United States Steel) for beams #1 and #2. Standard steel would have needed regular painting, which would be impossible because the deck makes the beams inaccessible. Cor-Ten Type A steel forms a protective surface layer of rust and doesn't need painting.

The western ends of beams #1 and #2 sit flush with the outside face of the west stub wall. The cantilevered eastern end of beam #1 was notched so that its bottom flange acted as a bearing point for one end of the diving-platform header. The cantilevered eastern end of beam #2 was cut at a 45° angle and notched to form an angled bearing point for beam #9.

Beam #9, the angled beam, is supported on the cantilevered ends of beams #4 and #2. In turn, one end of beam #9 cantilevers to catch the end of the diving platform header, and the other cantilevers to catch a step header. The western ends of beams #4 and #3 are cantilevered to catch the headers that form the triangular steps. These headers are supported by beams #7 and #8, which bear on 8-in. piers. These headers also tie into beams #5 and #6, which form the walkway to the deck.

In the step areas, Buchanan and his crew hung 2x12 pressure-treated joists flush with the beam system using joist hangers. Then they laid the deck framing, 2x8 pressure-treated joists, on top of and perpendicular to the wood and steel beams. This gave final shape to the deck.

The crew stapled metal insect screening on top of the deck joists under the gazebo location. Then they installed pressure-treated 1x6 decking over the entire deck and step framing. Next, 1x4 fir nosing was installed on the leading edges of the deck and the steps all the way around.

Retaining walls made of pressure-treated 6x6s were built within 1 in. of the entire deck perimeter. These walls hold in a 6 in. layer of crushed stone beneath the deck and hold back the fill and the topsoil that surround the deck. The tops of the retaining walls and the finished grading are 7 in. below the lowest deck step.

Gazebo construction—Except for the corners of the building, which are carried by exposed redwood 4x4s, the gazebo walls are framed with 2x4 studs supporting a continuous doubled 2x10 header and sheathed inside and out with ½-in. plywood, glued and nailed to prevent racking. The carpenters notched the continuous header at door locations to allow for head height.

The north wind. Carved by the designer and set into one of the pediments, this black-walnut sculpture depicts the site's prevailing wind and abundant four-leaf clover.

The eaves were built low and broad to provide a sense of cozy protection. For the hipped roof, I specified 2x12 rafters, not so much for their structural value as for their cross-sectional height on the low walls. This height lets the rafter tails project horizontally as far as possible without getting the eaves too low. Small pediments, framed over the doorways on the east and west facing roofs, allow extra height for doors.

We had planned for a copper standing-seam roof because of the turquoise patina that copper acquires. But in our area, acid rain reduces the finish of copper roofs to a mottled, dark brown to black. Instead we used a baked-enamel, standing-seam steel roof with a turquoise color (Ideal Roofing Co., 1418 Michael St., Ottawa, Ont., Canada K1B 3R2; 613-746-3206).

Cedar clapboards and redwood trim cover the exterior. In the early design stage, the gazebo had taken a classical form, so we added a 5/4x6 frieze board to suggest a simple entablature and wrapped stock cove moldings around the redwood columns to create capitals (photo right). Column bases were built of redwood 5/4x10 skirt boards. The screens are removable pine frames, contained by ½-in. quarter round screwed in place. Inside, the walls and the ceiling are finished in 1x4 V-groove redwood paneling.

I designed and carved a bas-relief sculpture out of black walnut and placed it in the west-facing roof pediment (photo above). The carving depicts a caricature of the north wind, the site's prevailing wind, and four-leaf clover, which are found in great abundance around the pond. The walnut was treated with five coats of marine-grade enamel on the back and five coats of marine-grade varnish on the face.

We originally planned to stain the gazebo the same gray-brown color as the main house. Once the redwood trim and the cedar clapboards were in place, however, it seemed a shame to lose the vitality of the natural wood. We left the redwood and the cedar untreated for a year, then applied a clear wood preservative.

Simple but elegant. Stock cove moldings, painted to match the nearby flora, evoke classical column capitals.

Treating the trim and the deck was another story: We chose colors from the landscape. Ajuga, a ground cover scattered throughout the gardens, has leaves that are green on one side and deep red on the other. The painters used these colors on all the trim and doors. The treated deck had to weather a year before it could absorb a stain. Then a soft, semitransparent gray stain was used to match the stone walls nearby.

The gazebo and the deck have been in place for over two years now. Our biggest concern has always been the stability of the unorthodox foundation. During the winter of 1989-90, temperatures frequently ranged from below 0° F to well above freezing. Many foundations suffered because of these temperature swings. Happily, the gazebo's foundation did not move. □

Alan Guazzoni is a designer in Stowe, Vt. Photos by Glenn Moody.

A Gazebo Showerhouse

Between dune and porch, a place to hose off the salt and the sand

by Charles Ballinger

Photo: Leroy Walker III

Along the barrier islands that make up the seashore resort towns of southern New Jersey, outdoor showers adjoin almost every house. These showers are primarily for outdoor use but often become the most popular shower "in the house." The combination of fresh air, ease of access and spaciousness make the experience very pleasurable.

A few years ago, my crew and I added a dining room to a large, oceanfront Queen Anne Victorian. The house is a stately old place, partially clad with fancy-cut cedar shingles, surrounded by complex porches and topped with a multitude of gables and dormers. On the northwest corner, a 3½-story tower dominates the house. The house had begun its slide into disrepair when our client bought it and took on the considerable task of breathing new life into the place. Before its restoration, the house was known locally as the "cobweb castle."

Years after the house was first built, one owner added an outdoor shower on the eastern wall. When we removed the wall to add the dining room, we had to remove the shower too—a loss that had to be remedied. Since the lot is large, we started to think about the new shower as a freestanding structure. Inspired by the tower in the main house, I proposed a shower resembling an old-time gazebo. For modesty's sake, the shower has shingled curtain walls (photo), but is open along the top for a view out. The walls stop about 1 ft. above grade to promote air flow.

Tying it to the ground—My first design problem concerned the exposed site. Fierce northeasterly winds blow in from the Atlantic every year, and because I didn't want to build the first Victorian space shuttle, I had to come up with a sturdy foundation.

Along these sandy islands, most homes are built on pressure-treated yellow pine pilings, usually water-jetted into the ground. The jet resembles a huge dental pick connected by a hose to a fire hydrant, and water under high pressure is used to erode a 15-ft. deep "quicksand" hole into which the pilings are dropped. As the jetted water dissipates, the sand locks onto the pilings.

This little gazebo didn't require pilings buried 15 ft. deep, but it did receive a base of pressure-treated posts. I used yellow pine 6x6s, treated with chromated copper arse-

Inspired by the tower on the neighboring Queen Anne Victorian, this outdoor shower on the New Jersey coast mimics the original house in form, color and materials.

nate (CCA), to make a hybrid piling/concrete slab foundation (drawing, below right). In plan, the gazebo is a hexagon, with a post at each corner. The posts are buried 4 ft. into the sand, and extend about 7 ft. above grade as columns to carry the roof. Each post has an L-shaped piece of ½-in. rebar driven into it, 3 in. below the surface of the slab. The lengths of rebar overlap one another where they converge at the center of the 6-in. thick slab. Encased in a monolithic pour, the columns have the substantial ballast of the concrete and the friction of embedded posts working to resist wind loads. And the troweled slab, crowned slightly for drainage, is a floor suitable for a shower.

There is a certain finality about concrete, and I realized the posts had to be set exactly right. The curtain walls had to be identical because they were going to be covered with patterned shingles. To avoid inaccuracies in the layout, I built a template to position the posts. I overlapped a couple of sheets of ⅜-in. plywood and screwed them together to make a full-size pattern of the post locations. Then I cut holes for the posts, and draped the entire affair over the excavated site.

Once I had the posts dropped into position and braced diagonally, I plumbed them with a few taps of the sledge. To lock the posts into alignment, I added the horizontal nailers for the plywood curtain walls.

As soon as we dismantled the template, we dug a trench for the water supply lines, which we encased in rubber sleeves where they would be covered with concrete. The pipes emerge at the base of the post that holds the showerhead. At the same time, we ran electric conduit to an exterior switch box now mounted by the door. Finally, we built 2x forms to cast a perfect hexagon with the posts 5 in. in from each corner.

Roof—After the pour, we took out the diagonal braces and built a six-sided scaffold using the posts as our inside uprights. We cut the posts to the proper height and notched their tops to receive a double 2x8 ring beam, as shown in detail drawing A.

The roof is unusual in that it is composed of six hip rafters and six common rafters. None of it was very common to me, so to simplify the rafter intersections I made a six-sided king post that would provide alignment and bearing for the tops of the rafters. To shape the king post, I set the blade of my table saw at 60° and took the corners off a length of 6x6, as shown in detail drawing B. The hip rafters meet the king post with simple plumb cuts; to get the common rafters to fit into the tight angle between the hips meant some tapering with the power plane.

For some visual interest on the inside of the structure, I let the king post hang down about 3 ft. to make a nailer for a "wagon wheel." The 2x2 spokes of the wheel are toe-

Charles Ballinger is a licensed general contractor specializing in remodeling.

nailed to the hip rafters. In addition to being an attractive visual element, the wheel makes a good base for a light fixture.

To add another element of interest to the ceiling, we used 1x6 roofer's decking (unsurfaced T&G yellow pine) instead of plywood. The narrow gaps between the 1x6s create lines of concentric hexagons that rise to the roof's peak. Eventually we spray-painted the ceiling white, giving it a clean, bright look.

When we built the dining-room addition on the main house, all the interior and exterior trim was custom milled, and I had ordered plenty. You don't want to end up 2 ft. shy on this stuff. Here the key phrase is "setup charges." Once the cutters are are set at the mill, you've stepped back in time to yesteryear. Of course, your checkbook won't confirm that, but I usually order extra because once the blades are dismantled, the molding is again extinct.

Since we wanted the gazebo trim to match the house trim, we ordered enough cove molding to dress up the gazebo fascia (detail drawing A). I positioned the cove molding so that it slightly lifts the bottom course of the roof decking. This gives the roof a slight flare at the bottom. To accentuate this curve, I installed a row of cedar shingles as a base for my starter row of asphalt shingles. Shingling

it was a straightforward but tedious job. And it takes a lot of hip and ridge shingles to do a roof like this.

Curtain walls—The curtain walls were sheathed and sided with Fancy Cut Shingles by Shakertown (P.O. Box 44, Winlock, Wash. 98596). We used arrow shingles and round bottom shingles, which form circles where they overlap (detail drawing C).

With shingles, I usually don't over-order fancy cuts. They are expensive, and very often I cut my own. I simply set my table saw to 5 in. and gang-cut groups of six to width until I've got a nice pile of them. Then I scribe a pattern on the top of the "six packs" and cut the butts to the desired profile with my saber saw. It goes quickly and it's a lot less expensive.

We finished the interior walls (and made the door) with V-joint T&G roughsawn cedar. When we renovated the main house, we finished the porch ceilings with this material, maintaining the aesthetic link. Along two walls are benches made of 5/4 by 6-in. cedar.

After spraying the little building with opaque stains to match the house, only one task remained. That crowning moment came when I climbed the roof and nailed a copper witch's hat to the peak. In this case, the golden spike had to be made of copper. □

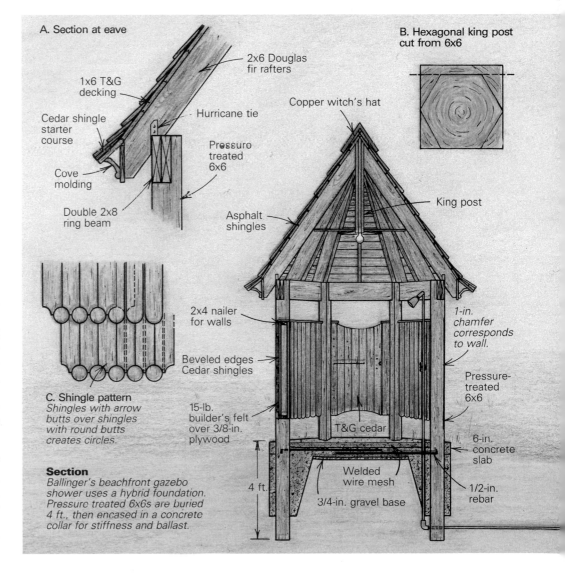

A. Section at eave

2x6 Douglas fir rafters

1x6 T&G decking

Cedar shingle starter course

Hurricane tie

Cove molding

Pressure treated 6x6

Double 2x8 ring beam

B. Hexagonal king post cut from 6x6

Copper witch's hat

Asphalt shingles

King post

2x4 nailer for walls

Beveled edges Cedar shingles

1-in. chamfer corresponds to wall.

Pressure-treated 6x6

C. Shingle pattern
Shingles with arrow butts over shingles with round butts creates circles.

15-lb. builder's felt over 3/8-in. plywood

T&G cedar

6-in. concrete slab

1/2-in. rebar

Welded wire mesh

3/4-in. gravel base

4 ft.

Section
Ballinger's beachfront gazebo shower uses a hybrid foundation. Pressure treated 6x6s are buried 4 ft., then encased in a concrete collar for stiffness and ballast.

Site-Built Footbridge

Using backcountry lamination techniques to span a creek in style

by Henry Smith

An arched bridge with benches. A pair of laminated redwood girders traverse this little ravine. At center span, a pair of built-in benches cantilever off the railing posts by way of half-lapped 4x4s.

There's nothing like a big event to focus your attention on some long-neglected job around the house. Maybe the new neighbors are dropping by for coffee, or the bank's appraiser is about to pass judgment on the home-equity loan. For me, it was a wedding day. In a couple of months, some friends were going to be married in the meadow below our house. To get to the meadow, the hundred or so guests would have to cross a small creek on our decaying, poorly constructed footbridge. The thought of launching a marriage with a bridge disaster was enough to get me in gear.

Trailside technology—On hiking trips in the Santa Cruz mountains near our home in northern California, I had admired the many small, finely crafted footbridges. Their decks were supported by laminated redwood girders that were built in place. The local park ranger sent me a drawing of one of their basic bridges. The drawing included the footing design and assembly techniques that I put to use building our bridge. I did change a couple of things, however.

I like the look of the arching footbridges that are often centerpieces of Japanese gardens. So I modified the straight girder the Park Service uses in its bridges by putting some curve into our bridge girders. Also, I decided to incorporate a pair of built-in benches in the railing (photo facing page). Located midspan, the benches provide a contemplative alcove to enjoy the creek, the natural setting and the company of friends.

Put the footings back from the bank—Our creek isn't very wide. But the banks are sandy clay held together by roots, and you can watch the bank erode before your eyes during a heavy downpour. So I chose to place the bridge footings about 8 ft. back from the erosion line. That put the two footings about 30 ft. apart.

The footings are 42 in. long and L-shaped in cross section (top drawing right). Both the vertical and the horizontal legs of the footings are 6 in. thick and 16 in. wide. The ends of the girders bear against the vertical legs, which oppose the horizontal thrust exerted by the arched girders. I placed three horizontal lengths of ½-in. rebar as well as vertical rebar Ls 6 in. o. c. to reinforce the concrete. Steel post bases cast into the footings anchor the ends of the girders. While building the footing forms, I made sure the footings ended up square and level to one another.

Overbuild the girder—I admit that I didn't have an engineer on this job. Instead, I made the girders so big that they *had* to be strong enough. Like the Park Service, I used green construction-heart redwood for the girders, but I used 2x8s instead of 2x6s. It's important to stagger the joints in a glued-up beam, so I made a chart on graph paper that served as a guideline for the glue up (bottom drawing right).

In my workshop I constructed a series of clamping stations to build up the beams with 8 in. of camber (top photo above). The two outboard stations were 28 ft. apart. Between them,

Girders under construction. The author shaped the girders by bending them across 2x4 clamping stations affixed to the shop walls. The outboard stations were weighted down with pier blocks to keep the 2x8s from lifting the stations.

A girder heads for home. Once the girders were cut to length, they were fitted with temporary handles and carried to their footings.

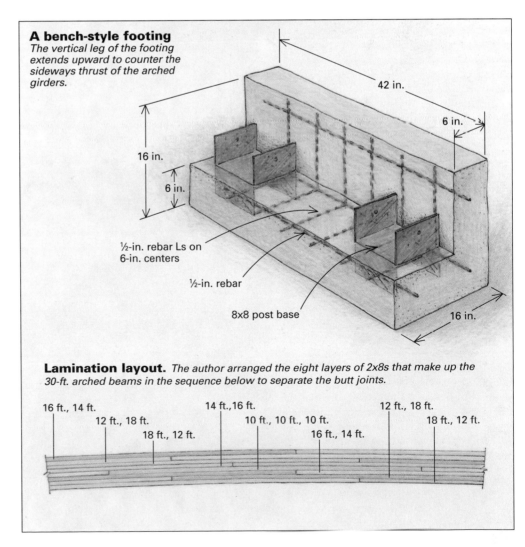

A bench-style footing
The vertical leg of the footing extends upward to counter the sideways thrust of the arched girders.

42 in.

6 in.

16 in.

6 in.

½-in. rebar Ls on 6-in. centers

½-in. rebar

8x8 post base

16 in.

Lamination layout. *The author arranged the eight layers of 2x8s that make up the 30-ft. arched beams in the sequence below to separate the butt joints.*

16 ft., 14 ft.

12 ft., 18 ft.

18 ft., 12 ft.

14 ft., 16 ft.

10 ft., 10 ft., 10 ft.

16 ft., 14 ft.

12 ft., 18 ft.

18 ft., 12 ft.

the center station was 8 in. taller. A couple of intermediate stations supported the stock at mid-curve. The stations consisted of 2x4s attached to the shop wall at a convenient working height and well-braced to allow for a considerable amount of weight and clamping force.

I joined the first two boards from underneath with a plywood splice plate to keep them aligned. Then I rolled Weldwood powdered resin glue onto the mating 2x8s, and pipe-clamped the 2x8s together beginning in the middle. I used the Weldwood glue because it's got a waterproof rating and because it has a long pot life.

I had to remove the clamps each time I added a new layer of boards. So before removing the clamps, I screwed the top board to the one below it with 2½-in. galvanized drywall screws on staggered 16-in. centers. The screws ensured that the glue joints would remain stationary as subsequent layers of 2x8s were added. Just to make sure the laminations stay put over the long haul, I ran ½-in. carriage bolts through the girders on staggered 2-ft. centers. Each girder took two people about four hours to build.

Cutting and setting the girders—As any builder knows, making an irreversible cut on an expensive beam can be a nerve-racking experience. And with a curved beam, "measure twice, cut once" doesn't apply. On the advice of a friend, I made a full-size pattern out of ½-in. plywood to mimic the shape of the girder. I stiffened the pattern with some 2x4s to keep it from being too floppy. Then, with the help of two assistants, I scribed the pattern in place to get the precise plumb and horizontal cuts.

The girders weighed about 700 lb. apiece, and they required eight men to move them into place. What's more, an arched girder wants to roll over as it is carried, so we nailed and clamped temporary handles to the girders (bottom photo, p. 117). The first girder dropped sweetly into place, with a gap of about ¹⁄₁₆ in. The second girder required a few passes with a power planer, and then it too fit.

Decking, posts and railings—I decked the bridge with redwood 2x6s by starting at the ends and working toward the middle. The ½-in. gaps between the boards promote quick drainage and keep leaves from clogging the gaps. When I neared the middle of the bridge, I ripped a few boards down to make sure that the deck boards fit comfortably and that the gaps stayed uniform.

I lag bolted 4x4 posts to the sides of the girders on 4-ft. centers. Then I clamped a flexible 2x ripping to the posts and adjusted the ripping to match the curve of the bridge at the right height for a railing. With the ripping as my guide, I marked the posts and cut them to length on the appropriate angle to receive the 2x6 railing.

With a lot of help from my friends, I finished the bridge the day before the wedding. We decorated the railings with balloons, crepe paper and flowers, and the wedding went off without a bridge collapse. □

Henry Smith lives in Santa Cruz, California. Photos by the author except where noted.

Another Way to Make an Arched Bridge

by Geoffrey Cole
As my 3-year-old son toddled onto the bridge, he shouted, "The bridge is falling." He was right. Rising water in our backyard creek had eroded the footings under the aging footbridge that connected our yard with our neighbors' yard. The bridge had sunk 18 in. Our neighbors and my family all wanted a new bridge, and I liked the idea of building it myself.

I began the bridge-building project by studying a book called *Bridges and* **Cupolas (Janet and Richard Strombeck, 1981, $12.45, Sun Designs, P. O. Box 6, Oconomowoc, Wis. 53066; 414-567-4255). The book contains sketches of 21 footbridges, eight covered bridges and 36 cupolas. I was immediately attracted to one of the covered bridges, but my grandiose plans never made it past the oversight committee: my wife and the neighbors. We finally decided on the 20-ft. arched wooden bridge that was called the "Potomac" in** *Bridges and Cupolas.*

Arched-beam bending form. Using a curved chalk mark on a concrete slab as his guide, the author made this bending form out of 2x4s. A strip of ¼-in. hardboard tacked to the tops of the risers defines the curve. The ¼-in. thick laminations were glued up in groups of five.

Stabilize the creek bank—Before investing the time and money to build a new bridge, I had to shore up the bank of the creek. To that end, I stacked unopened bags of premixed concrete in layers along the portion of the eroded bank. The bags were made of permeable paper, which allowed water to seep into the concrete. The concrete hardened, and in time, the bags ripped away, leaving a solid, durable wall that looks like a stack of concrete pillows.

The high cost of big, curved beams—The plan called for two 20-ft. arched beams, 3⅛-in. thick by 7½-in. deep. My local building-supply store had a source for the beams, but it wanted $980 apiece. That seemed like a lot of money, so I asked a woodworker friend if he was interested in the job. No thanks. So I made the beams myself and learned firsthand why a professional woodworker might not want to take on this tough job. It's a labor-intensive process.

I began by drawing the 22-ft. radius of the bottom of the beam on my garage floor. Then I made a bending form that matched the curve (photo facing page). The form consisted of a doubled 2x4 base with 2x4 risers spaced 6 in. o. c. The risers fit into ¾-in. deep dadoes in the base. I cut the top of each riser to match the angle of the full-size drawing. Then I screwed a strip of ¼-in. hardboard to the risers, and I waxed the hardboard to keep any misplaced glue from sticking to it.

I had a stack of clear pine siding left over from work on our house. Even though it isn't the most rot-resistant wood available, I decided to use the pine to fabricate my beams, reasoning that a good paint job and some annual maintenance would ensure a long and useful life span for the bridge.

The ¾-in. thick boards weren't easily bent with my low-tech tools. So I used my table saw and planer to rip them into supple, ¼-in. thick strips that were 3¼-in. wide by various lengths. In addition to being easier to bend, thin stock is less likely to spring back when the clamps have been removed.

Using a paint roller, I spread water-resistant yellow glue on the strips and then clamped them together in groups of five. I arranged the strips so that any butt joints in adjacent layers were separated by at least 3 ft. I let the strips cure overnight before removing the clamps. Then I added another five strips, and so on, until 28 strips were glued together.

I cut the finished beams to length and smoothed their sides with a hand-held power planer. After letting the glue cure for a couple of days, I finished the beams with a coat of primer and two coats of glossy white paint. Each beam weighed approximately 150 lb., and each beam required 672 linear ft. of pine, a half-gallon of glue and more than 20 hours of labor.

Steel brackets anchor the beams—The arched beams bear on concrete footings on each side of the creek. The footings, which are level to one another, are 6 ft. long, 2 ft. wide and nearly 4 ft. deep. A local welder fabricated the beam brackets out of ¼-in. steel. The brackets are connected to the footings by way of 18-in. anchor bolts welded to the bottoms of the brackets. U-shaped pockets in the brackets cradle the ends of the beams (photo top left), and ¾-in. bolts hold them in place.

I finished the bridge by adding a deck of pressure-treated 2x6s and stair treads of pressure-treated 2x12s. The treads sit atop stringers affixed to the inside faces of the arched beams. I got out the clamps again for the handrails. They're anchored to 4x4 posts bolted to the sides of the beams. I laminated the handrails in place out of four layers of ¼-in. thick pine.

I calculate the cost of the bridge's raw materials at a little more than $1,400 (that includes the wood for the beams), and I spent more than 100 hours putting the parts together. That's a lot of weekends for a project that isn't an essential piece of shelter. But the satisfaction of seeing the bridge each morning when my wife and I get up is payback enough (photo bottom left).

—When he isn't building structures in his garden, Geoffrey Cole is a neurosurgeon in Athens, Georgia. Photos by the author.

Twin arches span the creek. Held steady by bolts through steel brackets, the arches perch above the defunct bridge. The beam brackets, which are made of ¼-in. steel plate, were custom-made for the job.

A garden centerpiece. Painted white with a lattice border on the arched beams, the finished bridge unites neighboring yards. Pressure-treated 2x6s and 2x12s comprise the bridge decking and the treads at each end. The handrails were laminated in place from four layers of ¼-in. thick pine.

Outbuildings

Portability equals versatility. In a course taught by Steve Badanes, a frequent contributor to *Fine Homebuilding*, students at the Yestermorrow School designed and built this band shell for the town of Wakefield, Vermont. The structure is on skids so it can be moved. Photo by Barrie Fischer.

Water-tank gazebo. The windmill that used to sit on the roof of this building on Isle au Haut, Maine, pumped water into a holding tank below. Photo by Jefferson Kolle.

An artful overhang. Architect Peter Twombly extended the roof of this garage workshop in Hebron, New Hampshire, to shade the loft window. The overhang enhances the lines of the roof while it increases ventilation. Photo by Jefferson Kolle.

Backyard observatory beneath a clever disguise. With the roof closed, this building looks like any other rustic farm shack. However, John Phelps designed the roof in two halves that slide back on barn-door tracks, revealing an observatory within. Photo by Dick Mitchell.

Beautifully built barnette. Charles Miller built this simple storage shed entirely of lumber discarded by lumber stores. The walls are 2x6s on end, splined with rubber-gasket material slid into saw kerfs along each edge. Photo by Charles Miller.

An arched roof transforms a garden cottage from simple to special. Architect Robert Marx included a curve in the roofline of this Connecticut cottage, giving it a distinctive yet simple appearance. Photo by Robert Marx.

Gazebos and Belvederes:
Variations on a theme

Gazebo comes from the Middle English word for gaze, meaning to look upon intently. Belvedere is Italian for beautiful view. According to Webster's, both words refer to "a freestanding roofed structure, usually open on the sides." As the structures on these pages illustrate, this definition leaves a lot of room for interpretation.

Designer/builder: Robert Habiger, Albuquerque, N.M. Photo: John Olsen.

Picnic area outside Crawford, Neb. Photo: Brian Vanden Brink.

Facing page: Laud Holm Farm, Wells, Me. Photo: Brian Vanden Brink.

INDEX

The articles in this book originally appeared in *Fine Homebuilding* magazine. The date of first publication, issue number and page numbers for each article are given at right.